Torus Fibrations, Gerbes, and Duality

[redacted]

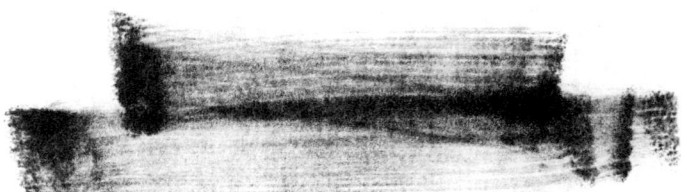

MEMOIRS
of the
American Mathematical Society

Number 901

Torus Fibrations, Gerbes, and Duality

Ron Donagi

Tony Pantev

(with an appendix by Dmitry Arinkin)

May 2008 • Volume 193 • Number 901 (first of 5 numbers) • ISSN 0065-9266

American Mathematical Society

Providence, Rhode Island

2000 *Mathematics Subject Classification.*
Primary 14A20, 14J32; Secondary 14A22, 14J27.

Library of Congress Cataloging-in-Publication Data

Donagi, Ron.
Torus fibrations, gerbes, and duality / Ron Donagi, Tony Pantev ; with an appendix by Dmitry Arinkin.
p. cm. — (Memoirs of the American Mathematical Society, ISSN 0065-9266 ; no. 901)
"May 2008, volume 193, number 901 (first of 5 numbers)."
Includes bibliographical references.
ISBN 978-0-8218-4092-4
1. Torus (Geometry) 2. Calabi-Yau manifolds. 3. Elliptic space. 4. Fourier transformations. I. Pantev, Tony, 1963– II. Title.

QA571.D66 2008
516.3′52—dc22 2008060002

Memoirs of the American Mathematical Society

This journal is devoted entirely to research in pure and applied mathematics.

Subscription information. The 2008 subscription begins with volume 191 and consists of six mailings, each containing one or more numbers. Subscription prices for 2008 are US$675 list, US$540 institutional member. A late charge of 10% of the subscription price will be imposed on orders received from nonmembers after January 1 of the subscription year. Subscribers outside the United States and India must pay a postage surcharge of US$38; subscribers in India must pay a postage surcharge of US$43. Expedited delivery to destinations in North America US$53; elsewhere US$130. Each number may be ordered separately; *please specify number* when ordering an individual number. For prices and titles of recently released numbers, see the New Publications sections of the *Notices of the American Mathematical Society.*

Back number information. For back issues see the *AMS Catalog of Publications.*

Subscriptions and orders should be addressed to the American Mathematical Society, P. O. Box 845904, Boston, MA 02284-5904, USA. *All orders must be accompanied by payment.* Other correspondence should be addressed to 201 Charles Street, Providence, RI 02904-2294, USA.

Memoirs of the American Mathematical Society (ISSN 0065-9266) is published bimonthly (each volume consisting usually of more than one number) by the American Mathematical Society at 201 Charles Street, Providence, RI 02904-2294, USA. Periodicals postage paid at Providence, RI. Postmaster: Send address changes to Memoirs, American Mathematical Society, 201 Charles Street, Providence, RI 02904-2294, USA.

This publication is indexed in *Science Citation Index*®, *SciSearch*®, *Research Alert*®, *CompuMath Citation Index*®, *Current Contents*®/*Physical, Chemical & Earth Sciences.*
Printed in the United States of America.

∞ The paper used in this book is acid-free and falls within the guidelines established to ensure permanence and durability.
Visit the AMS home page at `http://www.ams.org/`

10 9 8 7 6 5 4 3 2 1 13 12 11 10 09 08

Contents

Abstract

Let X be a smooth elliptic fibration over a smooth base B. Under mild assumptions, we establish a Fourier-Mukai equivalence between the derived categories of two objects, each of which is an $\mathcal{O}^{\times}$ gerbe over a genus one fibration which is a twisted form of X. The roles of the gerbe and the twist are interchanged by our duality. We state a general conjecture extending this to allow singular fibers, and we prove the conjecture when X is a surface. The duality extends to an action of the full modular group. This duality is related to the Strominger-Yau-Zaslow version of mirror symmetry, to twisted sheaves, and to non-commutative geometry.

Received by the editor January 20, 2005 and in a revised form on September 12, 2005.

2000 *Mathematics Subject Classification.* Primary 14A20, 14J32; Secondary 14A22,14J27.

Key words and phrases. elliptic fibrations, gerbes, Fourier-Mukai transforms, Calabi-Yau manifolds.

Ron Donagi was partially supported by NSF Grants DMS-0104354 and FRG-DMS-0139799.

Tony Pantev was partially supported by NSF grants DMS-0099715 and FRG-DMS-0139799, and an A.P.Sloan Research Fellowship.

CHAPTER 1

Introduction

1.1 Duality for elliptic fibrations

In this paper we are concerned with categories of sheaves on varieties fibered by genus one curves. For an elliptic fibration on X, by which we always mean a genus one fibration $\pi : X \to B$ admitting a holomorphic section $\sigma : B \to X$, there is by now a well understood theory of the Fourier-Mukai transform [**Muk81, BBRP98, Bri98, BM02**]. The basic result is:

Theorem [**BM02**] *Let* $X \underset{\sigma}{\overset{\pi}{\rightleftarrows}} B$ *be an elliptic fibration with smooth total space. Then the integral transform (Fourier-Mukai transform)*

$$\begin{aligned} \boldsymbol{FM} : D^b(X) &\longrightarrow D^b(X) \\ F &\longrightarrow Rp_{2*}(Lp_1^* F \otimes \mathscr{P}), \end{aligned}$$

induced by the Poincare sheaf $\mathscr{P} \to X \times_B X$*, is an auto-equivalence of the bounded derived category* $D^b(X)$ *of coherent sheaves on* X*.*

An important feature of $\boldsymbol{FM}$ is that it transforms geometric objects in an interesting way:

$$\left\{\begin{array}{l}\textit{Bundle data}\text{: vector bundles on } X\text{, semistable of}\\ \text{degree zero on the generic fiber of } \pi.\end{array}\right\}$$

$$\Big\updownarrow \boldsymbol{FM}$$

$$\left\{\begin{array}{l}\textit{Spectral data:}\text{ sheaves on } X \text{ with the numerics of}\\ \text{a line bundle on a `spectral' divisor } C \subset X\text{, finite}\\ \text{over } B.\end{array}\right\}$$

This spectral construction was used to study general compactifications of heterotic string theory and their moduli, and especially the duality with F-theory [**FMW97, BJPS97, Don97, AD98, FMW98,**

Don99, FMW99]. It was used also to construct special bundles on elliptic Calabi-Yau manifolds which lead to more-or-less realistic compactified theories [**DOPW01a, DOPW01b**].

1.2 Partial duality for genus one fibrations

In many applications (see e.g. [**DOPW01b, DOPW01a**]) one is also interested in constructing bundles on genus one fibrations $\pi : Y \to B$ which do not necessarily admit a section. From the viewpoint of the spectral construction one expects that vector bundles on Y should correspond to spectral data supported on a divisor $C \subset X := \overline{\mathrm{Pic}}^0(Y/B)$, where $\overline{\mathrm{Pic}}^0(Y/B)$ is the compactified relative Jacobian of $\pi : Y \to B$. However it is unrealistic to expect that this spectral data should again be a sheaf on C. One problem is that X is not a fine moduli space of objects on Y and so we do not have a Poincare sheaf on $Y \times_B X$. Another problem is that in the passage from Y to X we seem to be losing information. Indeed, there can be many different Y's with the same Jacobian X, and there is no obvious way to recover Y from spectral data consisting of a divisor on X and an ordinary sheaf supported on this divisor.

The resolution of this problem is suggested by string theory. Namely, in the transition from Y to X one should add an extra piece of data corresponding to the physicists' B-field. Mathematically the holomorphic version of this data is encoded in an $\mathcal{O}^\times$-gerbe on X. A detailed discussion of $\mathcal{O}^\times$-gerbes and their geometric properties can be found in [**Gir71, Bry93, Bre94, Hit01**] and in our section 1. A baby version of our result is that:

- Y determines a non-trivial $\mathcal{O}_X^\times$-gerbe ${}_YX$ on X.
- There is a gerby Fourier-Mukai transform exchanging

$$\left\{\begin{array}{l}\textit{Bundle data}\text{: vector bundles on } Y\text{, semistable of}\\ \text{degree zero on the generic fiber of } Y \to B.\end{array}\right\}$$

$$\Big\updownarrow \boldsymbol{FM}$$

$$\left\{\begin{array}{l}\textit{Spectral data:}\text{ sheaves on the gerbe } {}_YX \text{ with the}\\ \text{numerics of a line bundle on a `spectral' divisor}\\ ({}_YX) \times_X C\text{, with } C \subset X \text{ finite over } B.\end{array}\right\}$$

This statement is asymmetric - we only consider vector bundles on the variety Y, and the spectral data appears only for the gerbe

${}_YX$. The symmetry can be restored by extending this gerby spectral construction to a full Fourier-Mukai equivalence of derived categories.

One peculiarity of $\mathcal{O}^\times$-gerbes is that the categories of coherent sheaves on them, and hence also the derived categories, admit an orthogonal decomposition by subcategories labeled by the characters of $\mathcal{O}^\times$, i.e. by the integers (see section 1). For any $k \in \mathbb{Z}$, we will write $D^b_k({}_YX) \subset D^b({}_YX)$ for the k-th summand and we will call $D^b_k({}_YX)$ the derived category of weight k sheaves on ${}_YX$. It may be helpful to note here that when ${}_YX$ comes from an Azumaya algebra, the weight k corresponds to the central character of the action of this algebra.

Related partial dualities were considered previously in [**Căl02b, Căl02a**] in the context of Fourier-Mukai transforms and in [**DG02**] in the context of the spectral construction. A detailed analysis of the corresponding moduli spaces and the duality transformation between them was recently carried out, for the particular case of Hopf-like surfaces, in [**BM05, BM03**].

1.3 Main results: duality for genus one fibrations

For any elliptic fibration $X \underset{\sigma}{\overset{\pi}{\rightleftarrows}} B$, the twisted versions of X are parameterized by the analytic Tate-Shafarevich group $Ш_{an}(X)$ (see section 2). For a given $\beta \in Ш_{an}(X)$, let $\pi_\beta : X_\beta \to B$ denote the corresponding genus one fibration. On the other hand for any analytic space X, the analytic $\mathcal{O}^\times_X$-gerbes on X are parameterized by the analytic Brauer group $Br'_{an}(X) = H^2_{an}(\mathcal{O}^\times_X)$. For any $\alpha \in Br'_{an}(X)$ we denote the corresponding gerbe by ${}_\alpha X$ and the bounded derived category of coherent sheaves on ${}_\alpha X$ by $D^b({}_\alpha X)$. The latter decomposes naturally as the orthogonal direct sum of "pure weight" subcategories $D^b_k({}_YX)$ indexed by characters of $\mathcal{O}^\times_X$, i.e. by integers $k \in \mathbb{Z}$. For all $\alpha \in Br'_{an}(X)$ and all $k \in \mathbb{Z}$ there is a canonical equivalence $D^b_k({}_\alpha X) \cong D^b_1({}_{k\alpha}X)$. This follows immediately by comparing representations of ${}_\alpha X$ of pure weight k with representations of ${}_{k\alpha}X$ of pure weight 1 or by the appropriate idenitfications with the category of twisted sheaves on X (see section 1.1).

Consider first the case when X is a surface. We will see in section 3 that to any α, $\beta \in Ш_{an}(X)$ we can associate an $\mathcal{O}^\times$-gerbe ${}_\alpha X_\beta$ over X_β. The notation ${}_\alpha X_\beta$ generalizes our previous usage: when $\alpha = 0$ we get the trivial gerbe ${}_0X_\beta$ on X_β, and when $\beta = 0$ we get the gerbe ${}_\alpha X$ on $X = X_0$. Our main result in the case of surfaces, proved in chapter 4, is:

THEOREM A. *Let*

$$X \underset{\sigma}{\overset{\pi}{\rightleftarrows}} B = \mathbb{P}^1$$

be a non-isotrivial elliptic fibration on a smooth complex surface X*. Assume that* π *has* I_1 *fibers at worst. Let* $\alpha, \beta \in Ш_{an}(X)$ *be two elements such that* β *is torsion. Then there is an equivalence*

$$\boldsymbol{FM} : D^b_1({}_\alpha X_\beta) \to D^b_{-1}({}_\beta X_\alpha)$$

of the derived category of weight 1 *coherent sheaves on the gerbe* ${}_\alpha X_\beta$ *and the derived category of weight* (-1) *coherent sheaves on the gerbe* ${}_\beta X_\alpha$*. Equivalently* $\boldsymbol{FM}$ *can be thought of as an equivalence of the derived categories* $D^b_1({}_\alpha X_\beta)$ *and* $D^b_1({}_{-\beta} X_\alpha)$*.*

We will see in section 3 that for a higher dimensional X, the gerbes ${}_\alpha X_\beta$ may not make sense for arbitrary choices of $\alpha, \beta \in Ш_{an}(X)$. However in section 3 we show that there exists a natural pairing $\langle \bullet, \bullet \rangle$: $Ш_{an}(X) \otimes_{\mathbb{Z}} Ш_{an}(X) \to H^3_{an}(\mathcal{O}^\times_B)$ and that ${}_\alpha X_\beta$ can be defined whenever α, β are *complementary*, i.e. $\langle \alpha, \beta \rangle = 0$. The natural generalization of Theorem A is the following (see Conjecture 2.19):

Main Conjecture *For any complementary pair* $\alpha, \beta \in Ш_{an}(X)$*, there exists an equivalence*

$$D^b_1({}_\alpha X_\beta) \cong D^b_{-1}({}_\beta X_\alpha)$$

of the bounded derived categories of sheaves of weights ± 1 *on* ${}_\alpha X_\beta$ *and* ${}_\beta X_\alpha$ *respectively.*

We cannot prove this in full generality, mostly due to our inability to handle the general singular fibers. We are able to settle the conjecture in the non-singular case, under the somewhat more restrictive condition that α is m-divisible and β is m-torsion for some integer m. This of course implies that α, β are complementary, so ${}_\alpha X_\beta$ is well defined. For a smooth π our main result is:

THEOREM B. *Let*

$$X \underset{\sigma}{\overset{\pi}{\rightleftarrows}} B$$

be a smooth elliptic fibration on an algebraic variety X *over a smooth algebraic base* B*. Assume that* $Br'_{an}(B) = 0$*. Fix a positive integer* m

and let $\alpha, \beta \in \text{Ш}_{an}(X)$ be two elements such that α is m-divisible and β is m-torsion. Then there is an equivalence

$$\boldsymbol{FM} : D_1^b({}_\alpha X_\beta) \to D_{-1}^b({}_\beta X_\alpha)$$

of the derived category of weight 1 coherent sheaves on the gerbe ${}_\alpha X_\beta$ with the derived category of weight (-1) coherent sheaves on the gerbe ${}_\beta X_\alpha$. Equivalently $\boldsymbol{FM}$ can be thought of as an equivalence of $D_1^b({}_\alpha X_\beta)$ and $D_1^b({}_{-\beta} X_\alpha)$.

In fact, the proof of Theorem B is quite a bit easier than that of Theorem A, so we give it first, in chapter 3. The proof is based on the construction of two explicit presentations: the lifting presentation of ${}_\alpha X_\beta$, in section 1.1, and the extension presentation of ${}_\beta X_\alpha$, in section 1.2, together with the construction of an explicit Fourier-Mukai duality between them, in section 4.

Our two main theorems and their proofs have fairly straightforward analogues asserting the equivalence of the derived categories of quasi-coherent sheaves and, in the algebraic case, asserting the equivalence of appropriate categories of algebraic coherent sheaves. Indeed, if the class α happens to be torsion as well, then the spaces X_α and X_β are algebraic and the gerbes ${}_\alpha X_\beta$ and ${}_\beta X_\alpha$ are algebraic stacks in the sense of Artin. We will see in the proofs of the two theorems that the gerby Fourier-Mukai transform in this case will correspond to a kernel object which is algebraic and so will give rise to an equivalence of the derived categories of weight one algebraic coherent sheaves.

1.4 Duality for commutative group stacks

As was pointed out by Arinkin, our Theorem B (but not Theorem A) fits very naturally in the context of commutative group stacks (cgs). The $\mathcal{O}_X^\times$-gerbe ${}_\alpha X$ is a family of cgs over B which is an extension:

$$0 \to B\mathbb{G}_m \to {}_\alpha X_0 \to X \to 0$$

of X by the classifying stack of $\mathbb{G}_m$. The torsor X_β, on the other hand, is not a cgs over B; but it does determine one, namely the extension:

$$0 \to X \to \widetilde{X}_\beta \to \mathbb{Z} \to 0$$

of $\mathbb{Z}$ by X, where X_β is recovered as the inverse image of $1 \in \mathbb{Z}$. Similarly, the gerbe ${}_\alpha X_\beta$ which we construct, using either the lifting presentation (when β is m-torsion) or the extension presentation (when α is m-torsion), determines a cgs $\mathscr{X} = {}_\alpha \widetilde{X}_\beta$ which has a two-step filtration, with sub $B\mathbb{G}_m$, middle subquotient X, and quotient $\mathbb{Z}$. This can be considered either as an $\mathcal{O}_X^\times$-gerbe over X_β, or dually as a torsor over ${}_\alpha X$. In particular, the derived category $D^b({}_\alpha X_\beta)$ is graded by $\mathbb{Z} \times \mathbb{Z}$.

Now quite generally, such a cgs $\mathscr{X}$ has a dual cgs $\mathscr{X}^\vee$ which has a similar two-step filtration with the roles of the sub and the quotient interchanged. There is a Poincare sheaf $\mathscr{P}$ which is a biextension of $\mathscr{X}^\vee \times \mathscr{X}$ by $\mathbb{G}_m$, and it induces a Fourier-Mukai transform which is an equivalence of categories $D^b(\mathscr{X}^\vee) \simeq D^b(\mathscr{X})$ interchanging the two $\mathbb{Z}$ gradings. Our Theorem B can therefore be interpreted as saying that the cgs dual to ${}_\alpha\widetilde{X}_\beta$ is ${}_\beta\widetilde{X}_\alpha$; the previous version is recovered by restricting the equivalence to the piece of bidegree $(1,-1)$. This is explained in some more detail in the Appendix A which D. Arinkin kindly wrote for us.

This duality picture has a straighforward extension to more general cgs $\mathscr{X}$ over B: these are again endowed with a two step filtration $W_{-2}\mathscr{X} \subset W_{-1}\mathscr{X} \subset W_0\mathscr{X} = \mathscr{X}$, where $\mathrm{gr}_{-2} = W_{-2}\mathscr{X} = BT$ for some affine torus bundle $T \to B$, the middle subquotient gr_{-1} is some abelian scheme $A \to B$ and the last quotient gr_0 is some bundle $\Lambda \to B$ of finite rank free abelian groups over B. The dual cgs $\mathscr{X}^\vee$ is again the stack of homomorphisms $\underline{\mathrm{Hom}}_{\mathrm{cgs}}(\mathscr{X}, B\mathbb{G}_m)$ and one expects that in good cases the duality gives rise to an equivalence of the appropriate categories of representations.

Arinkin calls the cgs described in the previous paragraph 'Beilinson's one motives' since they were considered by Beilinson (unpublished) in the context of the theory of mixed motives. The cgs $\mathscr{X}$ are formally very similar to the classical one motives studied by Deligne in [**Del74**]. The one motives of [**Del74**] can be viewed either as certain mixed Hodge structures of level ≤ 1 or as cgs $\mathscr{M}$ defined over $\mathbb{C}$. As a commutative group stack, every Deligne's one motive $\mathscr{M}$ is equipped with a two step filtration $W_{-2}\mathscr{M} \subset W_{-1}\mathscr{M} \subset W_0\mathscr{M} = \mathscr{M}$, for which $\mathrm{gr}_{-2} = T$ for some affine torus T, $\mathrm{gr}_{-1} = A$ is a polarized abelian variety, and $\mathrm{gr}_0 = B\Lambda$ for some free abelian group Λ of finite rank. If we now look at families of Deligne's one motives defined over some base B we arrive at cgs over B which are of essentially the same shape as the Beilinson's one motives, but with the stackiness appearing at a different subquotient of the filtration. Furthermore, as explained in [**Del74**], the dual of a Deligne's one motive $\mathscr{M}$ is the cgs $\mathscr{M}^\vee := \underline{\mathrm{Hom}}_{\mathrm{cgs}}(\mathscr{M}, B\mathbb{G}_m)$, which is again a one motive of the same type with $\mathrm{gr}_{-2}\mathscr{M}^\vee = \mathrm{Hom}(\Lambda, \mathbb{G}_m)$, $\mathrm{gr}_{-1}\mathscr{M}^\vee = \widehat{A}$ (the dual abelian variety to A), and $\mathrm{gr}_0\mathscr{M}^\vee = B\,\mathrm{Hom}(T, \mathbb{G}_m)$.

In fact, we can view $\underline{\mathrm{Hom}}_{\mathrm{cgs}}(\bullet, B\mathbb{G}_m)$ as a transformation acting on commutative group stacks, which preserves the two natural families of Deligne's and Beilinson's one motives and induces a duality on each of these families. Moreover, since in both cases the duality is realized in terms of suitable biextensions of $\mathscr{X} \times \mathscr{X}^\vee$ and $\mathscr{M} \times \mathscr{M}^\vee$, one expects that the duality of cgs will give rise to an equivalence of the corresponding categories of representations of cgs. For the specific stacks ${}_\alpha\widetilde{X}_\beta$ this is precisely the content of our Theorem B.

1.5 The non-commutative aspect

Results having the general shape of Theorem A were anticipated in the physics literature. In fact, Ganor-Mihailov-Saulina have conjectured in [**GMS00**] that when Y is a genus one fibered $K3$ surface, there should exist a non-commutative deformation ${}_YX$ of $X = \overline{\mathrm{Pic}^0}(Y/B)$ and a categorical equivalence between instantons on ${}_YX$ and spectral data on Y. This is a special case of Theorem A.

This statement admits an intriguing interpretation in terms of non-commutative geometry, a topic currently of high interest to physicists [**NS98, KKO01**]. According to the general yoga of deformation quantization (see e.g. [**Kon01**]), any symplectic (or Poisson) structure on X is the first term in a non-commutative deformation of its structure sheaf. In a suitable algebro-geometric context, e.g. on a $K3$, a symplectic structure θ has three incarnations: as a real 2-form $\theta_{\mathbb{R}}$, a holomorphic 2-form $\theta^{2,0}$, or an antiholomorphic 2-form $\theta^{0,2}$. Then $\theta_{\mathbb{R}}$ determines a "non-commutative four-manifold", and $\theta^{2,0}$ determines a "non-commutative $K3$". The third incarnation, $\theta^{0,2}$, gives both the element X_θ in the Tate-Shafarevich group $\text{Ш}(X)$ and the $\mathcal{O}_X^\times$-gerbe ${}_\theta X$. In this sense, Theorem A can be viewed as an affirmative answer and a generalization of the [**GMS00**] conjecture.

1.6 Modified T-duality and the SYZ conjecture

The celebrated work of Strominger, Yau and Zaslow [**SYZ96**] interprets mirror symmetry of Calabi-Yau spaces in terms of special Lagrangian (SLAG) torus fibrations. If a CY manifold X (with "large complex struture") has mirror X', [**SYZ96**] conjecture the existence of fibrations $\pi : X \to B$ and $\pi' : X' \to B$ whose generic fibers are SLAG tori dual to each other: each parameterizes $U(1)$ flat connections on the other. In particular, each of these fibrations admits a SLAG zero-section, corresponding to the trivial connection on the dual fibers. The analogy with the theorem of [**BM02**] is clear: the SLAG torus fibration on the Calabi-Yau threefold replaces the elliptic fibration on the surface, and mirror symmetry (interchanging D-branes of type B with D-branes of type A) replaces the Fourier-Mukai transform (which interchanges vector bundles with spectral data).

Our work suggests that the SYZ conjecture should be extended to a SLAG analogue of Theorem A or of the Main Conjecture, in which the physical B-fields $\alpha \in H^2(X, \mathbb{R}/\mathbb{Z})$ play the role of our gerbes. This extension leads to an integrable system structure on the moduli space underlying mirror symmetry. We give an informal discussion of these matters in chapter 5, and we hope to return to them in future work.

1.7 Modularity

As often happens in physics, the Fourier-Mukai functor

$$\boldsymbol{FM} : D_1^b({}_\alpha X_\beta) \to D_1^b({}_{-\beta} X_\alpha)$$

is just one particular element of a whole collection of dualities. For simplicity consider only the case of a projective elliptic surface $\pi : X \to B$. In this case, our Fourier-Mukai duality works for any pair of elements $(\alpha, \beta) \in Ш(X) \times Ш(X)$ in the algebraic Tate-Shafarevich group. Thus the Fourier-Mukai functor corresponds to the action of the matrix

$$\begin{pmatrix} 0 & 1 \\ -1 & 0 \end{pmatrix} \in \mathrm{SL}(2, \mathbb{Z})$$

on the Cartesian square $Ш(X)^{\times 2}$ of the abelian group $Ш(X)$. Moreover, one can show (see e.g. section 3) that for surfaces the natural map $T_\beta : Ш(X) \to Br'(X_\beta)$, used to define our gerbes, has kernel generated by the element $\beta \in Ш(X)$. In particular, $T_\beta(\alpha + \beta) = T_\beta(\alpha)$ and so the gerbes ${}_{\alpha+\beta}X_\beta$ and ${}_\alpha X_\beta$ are isomorphic. A choice of such an isomorphism gives rise to an equivalence $D_1^b({}_{\alpha+\beta}X_\beta) \cong D_1^b({}_\alpha X_\beta)$ which corresponds to the action of the matrix

$$\begin{pmatrix} 1 & 1 \\ 0 & 1 \end{pmatrix} \in \mathrm{SL}(2, \mathbb{Z})$$

on $Ш(X)^{\times 2}$. Since these two matrices generate $\mathrm{SL}(2, \mathbb{Z})$, it will be very interesting to investigate which braid group extension of $\mathrm{SL}(2, \mathbb{Z})$ acts on $\coprod_{\alpha, \beta \in Ш(X)} D_1^b({}_\alpha X_\beta)$, lifting the action on $Ш(X)^{\times 2}$. In the case when the Mordell-Weil group of X is trivial we expect this extension to be central and to be related to the extensions appearing in [**Pol96, Pol02**], [**Orl02**] and [**ST01**]. We do not discuss this question here but hope to return to it in a future work.

1.8 Twisted sheaves

Another context in which Theorem A turns out to be relevant is the theory of twisted sheaves on a complex space which admits a genus one fibration. Recall [**Căl00**] that for any $\mathcal{O}^{\times}$-valued Čech 2-cocycle α on a complex space X, one can consider the abelian category of α-twisted sheaves on X and its derived category $D^b(X, \alpha)$. By definition, an α-twisted sheaf on X is a collection of coherent sheaves defined over open sets in X, together with gluing data on overlaps which satisfy the α-twisted cocycle condition on triple overlaps (see [**Căl00**] or our section 1.1 for details). Refininig the open covering or changing the cocycle by a coboundary results in an equivalent category of twisted sheaves. Twisted sheaves on Calabi-Yau manifolds, and in particular on genus one fibered Calabi-Yau manifolds, were recently studied by A.Căldăraru [**Căl00, Căl02b**]. In particular, he observed [**Căl02a**] that in the case of a K3 surface, the derived category of α-twisted sheaves possesses certain natural Fourier-Mukai partners. The starting point of his analysis is the observation that if X is a smooth projective K3 surface, then every element $\alpha \in H^2_{et}(X, \mathcal{O}^{\times})$ can be interpreted as a homomorphism $\alpha : \boldsymbol{T}_X \to \mathbb{Q}/\mathbb{Z}$, where $\boldsymbol{T}_X$ denotes the transcendental lattice of X (see [**Căl02a**] or section 1.1). This interpretation suggests the following:

Căldăraru's Conjecture Let X and Y be two projective K3 surfaces and let $\alpha \in H^2_{et}(X, \mathcal{O}^{\times})$ and $\beta \in H^2_{et}(Y, \mathcal{O}^{\times})$. Then the derived categories $D^b(X, \alpha)$ and $D^b(Y, \beta)$ are equivalent as triangulated categories iff the lattices $\ker(\alpha) \subset \boldsymbol{T}_X$ and $\ker(\beta) \subset \boldsymbol{T}_Y$ are Hodge isometric.

When both α and β are zero, the conjecture is true in view of a theorem of D.Orlov [**Orl97**] asserting that two smooth projective K3 surfaces have equivalent derived categories iff their transcendental lattices are Hodge isometric. This has been extended by Căldăraru, who used Mukai's quasi-universal sheaves for non-fine moduli spaces of sheaves on K3 surfaces to deduce that the conjecture holds whenever one of the classes, say β, is trivial. The algebraic case of our Theorem A proves Căldăraru's Conjecture in a series of new cases, with both α and β non- zero.

Indeed, if $\pi : X \to B$ is an elliptic K3 surface and if $\alpha, \beta \in Ш(X)$ are two elements in the algebraic Tate-Shafarevich group, then the natural identification $Ш(X) = H^2_{et}(X, \mathcal{O}^\times)$ coming from the Leray spectral sequence allows us to view both α and β as homomorphisms $\boldsymbol{T}_X \to \mathbb{Q}/\mathbb{Z}$. Using this interpretation one checks immediately that the transcendental lattices of the K3 surfaces X, X_α and X_β satisfy $\boldsymbol{T}_{X_\alpha} = \ker(\alpha) \subset \boldsymbol{T}_X$ and $\boldsymbol{T}_{X_\beta} = \ker(\beta) \subset \boldsymbol{T}_X$, where it is understood that all the equalities are Hodge isometries. Let $T_\beta(\alpha) \in Br'_{an}(X_\beta) = H^2_{an}(X_\beta, \mathcal{O}^\times)$ denote the class of the gerbe ${}_\alpha X_\beta$. Assuming that α and β are in general position in $H^2_{et}(X, \mathcal{O}^\times)$, i.e. that the cyclic subgroups generated by α and β intersect only at zero, we have natural identifications of Hodge lattices:

$$\ker\left[\boldsymbol{T}_{X_\alpha} \xrightarrow{T_\alpha(\beta)} \mathbb{Q}/\mathbb{Z}\right] = \ker(\alpha) \cap \ker(\beta) \subset \boldsymbol{T}_X$$

$$\ker\left[\boldsymbol{T}_{X_\beta} \xrightarrow{T_\beta(\alpha)} \mathbb{Q}/\mathbb{Z}\right] = \ker(\alpha) \cap \ker(\beta) \subset \boldsymbol{T}_X.$$

In other words - the hypothesis of Theorem A implies the hypothesis of Căldăraru's conjecture. Combined with the remark that $D^b(X_\alpha, T_\alpha(\beta)) \cong D^b_1({}_\alpha X_\beta)$ and $D^b(X_\beta, -T_\beta(\alpha)) \cong D^b_{-1}({}_\beta X_\alpha)$, this shows that Theorem A implies an interesting new case of Căldăraru's conjecture (see Corollary 4.6 for a slightly more general statement). Note that the condition (required in the statement of Theorem A) that β should be torsion, is vacuous in this case, since for a smooth projective surface X, both the cohomological Brauer group $H^2_{et}(X, \mathcal{O}^\times)$ and the Tate-Shafarevich group $Ш(X)$ are torsion groups.

Recently a slightly modified version of Căldăraru's conjecture was proven by Huybrechts and Stellari [**HSt04a, HSt04b**]. Their formulation of the conjecture requires a Hodge isometry between twisted Mukai lattices rather than kernels of Brauer classes. It is not hard to check that in the elliptic setting of [**HSt04b**, Section 1(vi)] the hypothesis of Theorem A also implies the hypothesis of the [**HSt04b**, Theorem 0.1]. Thus Theorem A is compatible with the Huybrechts-Stellari result in the special case of projective elliptic $K3$ surfaces.

The paper is organized as follows. In chapter 2 we recall some standard facts about the geometry of $\mathcal{O}^\times$ gerbes and genus one fibrations. We also derive the compatibility condition between two Tate-Shafarevich classes and state a general conjecture on the equivalence of derived categories for gerbes over genus one fibrations. In chapter 3 we introduce the main characters appearing in the proofs of the two theorems stated above. Working in the setup of Theorem B, we define two geometric presentations - the lifting and extension presentations

- for the gerbes ${}_{\alpha}X_{\beta}$ and ${}_{\beta}X_{\alpha}$. Furthermore, we construct an integral transform between the corresponding atlases. We show that this integral transform sends descent data to descent data and gives rise to an equivalence of the derived categories of the gerbes, thus proving Theorem B. Chapter 4 deals with the case of surfaces. We show how, in the case of a surface, one can extend the lifting and extension presentations across the singular fibers and produce a Fourier-Mukai transform between the corresponding gerbes. We again check that this transform is an equivalence, which proves Theorem A. Finally, in chapter 5 we discuss the analogy between algebraic gerbes over genus one fibrations and flat gerbes over SLAG 3-torus fibrations on Calabi-Yau threefolds. We describe a conjectural picture which amends the Strominger-Yau-Zaslow version of mirror symmetry to incorporate non-trivial B-fields on both sides of the mirror correspondence.

Acknowledgments: We would like to thank D.Arinkin, A.Căldăraru, M.Gross, A.Kresch and D.Orlov for insightful discussions on the subject of this paper. We would also like to thank KITP and MSRI for providing a stimulating research environment during certain stages of the preparation of this work.

CHAPTER 2

The Brauer group and the Tate-Shafarevich group

We need some basic facts relating elements of the Brauer group to elements of the Tate-Shafarevich group of an elliptic fibration. We discuss $\mathscr{O}^\times$-gerbes and the Brauer groups which classify them in section 1, then genus-1 fibrations and the Tate-Shafarevich group which classifies them, in section 2. For an elliptic fibration there is a simple, direct relation between these two groups. The extension to genus-1 fibrations though is more delicate, and is defined only when a certain alternating pairing vanishes. This is discussed in section 3.

1. Brauer groups and $\mathscr{O}^\times$-gerbes

In this section we review the notions of $\mathscr{O}^\times$-gerbe and presentation, and discuss the relationship between $\mathscr{O}^\times$-gerbes and elements in the Brauer group.

1.1. $\mathscr{H}$-gerbes. Let $\mathscr{H}$ be a sheaf of abelian groups on a topological space (or a site) X. The case of main interest for us is when $(X, \mathscr{O}_X)$ is a ringed space and $\mathscr{H} = \mathscr{O}_X^\times$ is the sheaf of invertible elements in the structure sheaf. In fact, most of the time we will have $\mathscr{H} = \mathscr{O}_X^\times$ in either the etale or the analytic topologies on a complex scheme (or an algebraic or analytic space) X. In chapter 5 we will be interested also in the case when $\mathscr{H}$ is the sheaf of germs of smooth maps from a C^∞ manifold X to the circle S^1.

An $\mathscr{H}$-gerbe on X is a global structure on X which "locally looks like the quotient of X by the trivial action of $\mathscr{H}$". More precisely "the quotient of X by the trivial action of $\mathscr{H}$" is the classifying object $B\mathscr{H}$. For example, in case $\mathscr{H}$ is the sheaf of holomorphic maps from X to a fixed group H, $B\mathscr{H}$ is the sheaf of sections of $X \times BH$ over X, where BH is the classifying space of H. In the general case, $B\mathscr{H}$ can be interpreted either as a topological space over X (defined up to homotopy), or as a stack in groupoids over X (see [**LMB00**, §3] for the definition). We adopt the second approach and treat $B\mathscr{H}$ as a stack (='sheaf of categories'): over any open set V, the objects of $B\mathscr{H}(V)$ are the $\mathscr{H}$-torsors on V and the morphisms are the isomorphisms of torsors. In particular, the automorphisms of the trivial torsor $\mathbf{1}_V$ are given by elements in $\mathscr{H}(V)$. Note that $B\mathscr{H}$ is in fact a commutative group stack over X with a group structure given by convolution of $\mathscr{H}$-torsors. Explicitly, for any two $\mathscr{H}$-torsors A' and A'' over V the convolution $A' \otimes A''$ is defined as the $\mathscr{H}$-torsor $(A' \times A'')/\ker(\boldsymbol{m}_{\mathscr{H}})$, where $\boldsymbol{m}_{\mathscr{H}} : \mathscr{H} \times \mathscr{H} \to \mathscr{H}$ is the multiplication map.

DEFINITION 2.1. An $\mathscr{H}$-gerbe on X is a $B\mathscr{H}$ torsor, i.e. a stack of groupoids ${}_{\alpha}X$ over X, which is equipped with a principal homogeneous action of $B\mathscr{H}$.

REMARK 2.2. • Explicitly, a stack of groupoids ${}_{\alpha}X \to X$ is an $\mathscr{H}$-gerbe if for any open $V \subset X$ and any object s of ${}_{\alpha}X(V)$ we have chosen isomorphisms $\mathscr{H}(V) \cong \mathrm{Aut}_{{}_{\alpha}X(V)}(s)$, compatible with pullbacks.

• In the literature [**Gir71**], [**Bre90, Bre94**], one encounters a more general notion of an $\mathscr{H}$-gerbe, namely - a stack $\mathscr{T}$ of groupoids on X, which is locally isomorphic to $B\mathscr{H}$. These more general gerbes are classified by the first cohomology of X with coefficients in the 1-truncated simplicial abelian group $\mathscr{H} \to \mathrm{Aut}(\mathscr{H})$ [**Bre90**]. They are intimately related to the forms of $\mathscr{H}$, i.e. to sheaves of groups on X which are only locally isomorphic to $\mathscr{H}$. This relatishionship is based on the identification $\mathrm{Out}(\mathscr{H}) = \underline{\mathrm{Aut}}_X(B\mathscr{H})$: to any $\mathscr{T} \to X$, which is an $\mathscr{H}$-gerbe in this more general sense, one naturally associates an $\mathrm{Out}(\mathscr{H})$-torsor $\mathrm{band}(\mathscr{T}) := \underline{\mathrm{Isom}}_X(\mathscr{T}, B\mathscr{H})$ - the band of the gerbe $\mathscr{T}$ [**Gir71**], [**Bre90**]. A gerbe $\mathscr{T}$ is said to be banded by $\mathscr{H}$ if it is equipped with a trivialization of the torsor $\mathrm{band}(\mathscr{T})$. When $\mathscr{H}$ is abelian, this condition is equivalent to requiring that for any open V and any $s \in \mathscr{T}$ we have chosen isomorphisms $\mathscr{H}(V) \cong \mathrm{Aut}_{\mathscr{T}}(s)$ in a way compatible with pullbacks. In other words, the more restrictive notion of an $\mathscr{H}$-gerbe that we have adopted in this paper is the same as the standard notion of an $\mathscr{H}$-banded gerbe (at least for an abelian $\mathscr{H}$). We will casually ignore this distinction and will call all our gerbes simply $\mathscr{H}$-gerbes.

In case $\mathscr{H} = \mathcal{O}_X^{\times}$ (in the relevant topology), the classifying stack $B\mathcal{O}_X^{\times}$ is the sheaf of Picard categories $\mathscr{P}ic(X)$: for an open U, the objects of $\mathscr{P}ic(X)(U)$ are by definition the line bundles on U, and for two objects $L, M \in \mathrm{ob}(\mathscr{P}ic(X)(U))$ the set $\mathrm{Hom}_{\mathscr{P}ic(X)}(L, M)$ is defined to be $\mathrm{Isom}(L, M)$. An $\mathcal{O}_X^{\times}$ gerbe ${}_{\alpha}X$ assigns to each open U a $\mathscr{P}ic(X)(U)$-torsor, denoted $\mathscr{P}ic_{\alpha}(U)$, with a compatibility of the assignments to different U's. We can thus think of a section of an $\mathcal{O}_X^{\times}$-gerbe as a twisting of the notion of a line bundle on X. More generally the sections in an $\mathscr{H}$-gerbe are twistings of the notion of an $\mathscr{H}$-torsor on X: simply replace in the previous discussion each appearance of $\mathscr{P}ic$ with $\mathscr{T}ors^{\mathscr{H}}$ - the group of $\mathscr{H}$-torsors. This interpretation suggests that the group classifying $\mathscr{H}$-gerbes should be $H^2(X, \mathscr{H})$. When $\mathscr{H}$ is abelian this statement can be made precise via the standard cohomological machinery [**Mil80**, IV.2] or [**Gir71**], [**Bre90**] (but keep in mind that our $\mathscr{H}$- gerbes are the $\mathscr{H}$-banded gerbes of loc. cit.).

In more down to earth terms the interpretation of the elements in $H^2(X, \mathscr{H})$ as equivalence classes of gerbes can be seen as follows. Assume that we are in the good situation when the cohomology of $\mathscr{H}$ can be computed in Čech terms. Let $\{\alpha_{ijk}\}$ be an $\mathscr{H}$-valued Čech 2-cocycle w.r.t. an open cover $\{U_i\}$ of X. An object L of $\mathscr{T}ors^{\mathscr{H}}_{\alpha}$ is defined to be an assignment of an $\mathscr{H}(U_i)$-torsor $L(U_i)$ to each U_i, together with transition functions

$$g_{ij} : L(U_i) \otimes_{\mathscr{H}(U_i)} \mathscr{H}(U_{ij}) \widetilde{\rightarrow} L(U_j) \otimes_{\mathscr{H}(U_j)} \mathscr{H}(U_{ij})$$

satisfying the twisted cocycle condition:

$$g_{ij} \circ g_{jk} \circ g_{ki} = \alpha_{ijk}$$

on triple intersections. A morphism between two α-twisted $\mathscr{H}$-torsors L' and L'' is given by a compatible collection of isomorphisms $L'(U_i) \widetilde{\rightarrow} L''(U_i)$.

Similarly we define the category $\mathscr{T}ors^{\mathscr{H}}_{\alpha}(U)$ for any open U. The resulting sheaf of categories (=stack) $\mathscr{T}ors^{\mathscr{H}}_{\alpha}$ on X is by definition a torsor over $B\mathscr{H} = \mathscr{T}ors^{\mathscr{H}}_{\mathbf{1}}$, i.e. an $\mathscr{H}$-gerbe, which we denote by ${}_{\alpha}X$. Clearly two cocylces which represent the same cohomology class in $\check{H}^2(X, \mathscr{H})$ define isomorphic gerbes. Conversely, if sheaf cohomology on X can be computed in Čech terms, any $\mathscr{H}$-gerbe arises this way from some α w.r.t. a sufficiently refined cover [**Gir71**], [**Bre90, Bre94**].

Notation: • Given an $\mathscr{H}$-gerbe $\mathscr{T}$ over X we write $[\mathscr{T}] \in H^2(X, \mathscr{H})$ for the element that classifies it.

• The base space X for an $\mathscr{H}$-gerbe $\mathscr{T} \to X$ is called the *coarse moduli space* of $\mathscr{T}$. This terminology reflects the fact that X represents the sheaf of sets $\pi_0(\mathscr{T})$, i.e. the sheaf of isomorphism classes of sections in $\mathscr{T}$.

Basic construction: Starting with an algebraic (or analytic) space X and a short exact sequence of sheaves of groups

$$1 \to \mathscr{H} \to \mathscr{G} \to \mathscr{K} \to 1,$$

with $\mathscr{H}$-abelian, we get a coboundary map

$$\delta : H^1(X,\mathscr{K}) \to H^2(X,\mathscr{H}).$$

This admits the following lift on the level of torsors and gerbes: a $\mathscr{K}$-torsor $\mathscr{C}$ with class $[\mathscr{C}] \in H^1(X,\mathscr{K})$ determines an $\mathscr{H}$-gerbe $\delta(\mathscr{C})$ with class $\delta([\mathscr{C}]) \in H^2(X,\mathscr{H})$. Explicitly, for an open U, $\delta(\mathscr{C})(U)$ is the category of pairs $(\mathscr{D},\iota)$ where $\mathscr{D}$ is a $\mathscr{G}$-torsor on U and $\iota : \mathscr{D}\times_{\mathscr{G}}\mathscr{K} \to \mathscr{C}$ is an isomorphism of $\mathscr{K}$-torsors on U.

A familiar special case involves the sequence

$$1 \to \mathcal{O}_X^\times \to GL_n(\mathcal{O}_X) \to \mathbb{P}GL_n(\mathcal{O}_X) \to 1.$$

It says that every projective bundle on X gives rise to an $\mathcal{O}_X^\times$-gerbe which is trivial if and only if the projective bundle is a projectivization of a vector bundle.

If $(X,\mathcal{O}_X)$ is a nice ringed space for which cohomology can be computed in Čech terms, then the choice of $\alpha \in H^2(X,\mathcal{O}_X^\times)$ gives rise to the notion of α-twisted sheaves on X. More precisely, let $\mathfrak{U} = \{U_i\}$ be an open cover of X (in the topology under consideration) and let

$$\underline{\alpha} = \{\alpha_{ijk}\} \in \check{C}^2(\mathfrak{U},\mathcal{O}_X^\times)$$

be a 2-cocycle representing $\alpha \in H^2(X,\mathcal{O}_X)$. One defines an $\underline{\alpha}$-twisted sheaf on X as a collection $\{F_i\}$ of sheaves $F_i \to U_i$ of $\mathcal{O}_X$-modules, together with a collection of gluing isomorphisms

$$\varphi_{ij} : F_{j|U_{ij}} \xrightarrow{\cong} F_{i|U_{ij}}$$

satisfying $\varphi_{ii} = \mathrm{id}$, $\varphi_{ij} = \varphi_{ji}^{-1}$, and $\varphi_{ij}\circ\varphi_{jk}\circ\varphi_{jk} : F_{i|U_{ijk}} \to F_{i|U_{ijk}}$ is given by multiplication by α_{ijk}. Given two $\underline{\alpha}$ twisted sheaves $\boldsymbol{F} = \{F_i,\varphi_{ij}\}$ and $\boldsymbol{G} = \{G_i,\gamma_{ij}\}$ we define a homomorphism $\boldsymbol{f} : \boldsymbol{F} \to \boldsymbol{G}$ to be a collection $\boldsymbol{f} = \{f_i\}$ of sheaf morphisms $f_i : F_i \to G_i$ satisfying $f_i\circ\varphi_{ij} = \gamma_{ij}\circ f_j$. Composition is defined in an obvious way and so we obtain a category of $\underline{\alpha}$-twisted sheaves, which depends both on the cover $\mathfrak{U}$ and on the cocycle $\underline{\alpha}$. It can be checked [**Căl00**, Section 1.2] that the operations of passing to a refinement $\mathfrak{U}'$ of $\mathfrak{U}$ and of replacing $\underline{\alpha}$ by a cohomologous cocycle $\underline{\alpha}'$, give rise to an equivalent category of $\underline{\alpha}'$-twisted sheaves. Thus for any $\alpha \in H^2(X,\mathcal{O}_X^\times)$ we get category $(\mathcal{O}_X,\alpha)$-mod of α-twisted sheaves on X (defined only up to a non-canonical equivalence). An α-sheaf $\boldsymbol{F}$ on X is called quasi-coherent (respectively coherent) if each F_i is quasi-coherent (respectively coherent). We will write $\mathrm{QCoh}(X,\alpha)$ and $\mathrm{Coh}(X,\alpha)$ for the categories of quasi-coherent and coherent α-twisted sheaves. Note that $(\mathcal{O}_X,\alpha)$-mod, $\mathrm{QCoh}(X,\alpha)$ and $\mathrm{Coh}(X,\alpha)$ are all abelian categories.

More intrinsically the α-twisted sheaves on X can be interpreted as weight one sheaves on ${}_\alpha X$, where a sheaf on ${}_\alpha X$ is understood as a representation of the sheaf of groupoids ${}_\alpha X \to X$. To spell what this means, let us denote by $\mathscr{Q}\operatorname{Coh}_X$ the stack of quasicoherent sheaves on the space X. Let $\mathscr{X} \to X$ be any fibered category over X. Recall (see e.g. [**Del90**, Section 3.3] or [**LMB00**, Definition 13.3.3]) that a *representation* of $\mathscr{X}$ is a morphism $F : \mathscr{X} \to \mathscr{Q}\operatorname{Coh}_X$ of fibered categories defined over X. Explicitly this means that for any algebraic (or analytic) space $T \to X$ we are given a finctor $F_T : \mathscr{X}(T) \to \operatorname{QCoh}(T)$ so that F_T is compatible with base changes.

In particular, if ${}_\alpha X \to X$ is an $\mathcal{O}_X^\times$-gerbe, then a representation of ${}_\alpha X$ is a X-functor $F : {}_\alpha X \to \mathscr{Q}\operatorname{Coh}_X$. Given an integer n we say that F is a *pure* ${}_\alpha X$*-representation of weight* n if for any open $U \subset X$ and any section $L \in {}_\alpha X(U)$ the natural sheaf homomorphism $\underline{\operatorname{Aut}}_U(L) \to \underline{\operatorname{Aut}}_U(F(L))$ induced by F factors as

$$\begin{array}{ccc} \underline{\operatorname{Aut}}_U(L) & \longrightarrow & \underline{\operatorname{Aut}}_U(F(L)) \\ \Big\| & & \uparrow \\ \mathcal{O}_U^\times & \xrightarrow{(\bullet)^n} & \mathcal{O}_U^\times \end{array}$$

where in the bottom row the map is the raising into power n. It is instructive to point out that when we are dealing with the trivial gerbe ${}_0 X$ on X, a representation of ${}_0 X$ is nothing but a quasicoherent sheaf F on X equipped with a direct sum decomposition $F = \oplus_{n\in\mathbb{Z}} F_n$ into quasicoherent sheaves F_n so that a locally defined function $f \in \mathcal{O}_X$ acts on F as multiplication by f^n on F_n. The reader can check as an exercise that the category of representations of ${}_\alpha X$ of pure weight one is equivalent to the category $\operatorname{QCoh}(X, \alpha)$ and that the category of representations of ${}_\alpha X$ of pure weight n is equivalent to the category $\operatorname{QCoh}(X, n\alpha)$.

1.2. Geometric gerbes and their presentations. In this section we recall a more geometric approach to $\mathscr{H}$-gerbes which involves gluing of certain good local models. This exploits the standard idea that various geometric objects can be conveniently presented in terms of an atlas modulo certain gluing relations on it. For example, for a manifold X, an atlas U can be taken to be the disjoint union $U = \coprod_i U_i$ of coordinate charts, and the gluing can be specified by the closed subset of relations

$$R := U \times_X U \subset U \times U,$$

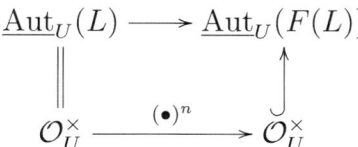

which comes together with two maps $\boldsymbol{s}, \boldsymbol{t} : R \to U$ (corresponding to the two projections of $U \times U$ onto U) each of which is a local diffeomorphism. Analogously, presentations can be used to define schemes, algebraic spaces and analytic spaces.

Formally, a *presentation by objects in a (fibered) category $\mathscr{C}$ (or a groupoid in $\mathscr{C}$)* consists of the following data:

(atlas)	an object U of $\mathscr{C}$
(relations)	an object R of $\mathscr{C}$
(source-target maps)	$R \overset{\boldsymbol{s}}{\underset{\boldsymbol{t}}{\rightrightarrows}} U$
(composition map)	$R \times_U R \xrightarrow{\boldsymbol{m}} R$
(inversion map)	$R \xrightarrow{\boldsymbol{i}} R$
(identity map)	$U \xrightarrow{\boldsymbol{e}} R.$

These data are subject to the obvious analogues of the group axioms, applied to the maps $\boldsymbol{m}$, $\boldsymbol{i}$ and $\boldsymbol{e}$.

Note that any morphism $\gamma : U \to X$ in $\mathscr{C}$ determines a presentation $(\mathfrak{R}, U, \mathfrak{m}, \mathfrak{i}, \mathfrak{e})$ in $\mathscr{C}$, where: $\mathfrak{R} := U \times_X U$; the maps $\mathfrak{s}$, $\mathfrak{t}$ are the two projections; the composition map $\mathfrak{m}$ sends $(a,b) \times (b,c)$ to (a,c), $\mathfrak{i}$ sends (a,b) to (b,a); and $\mathfrak{e}$ is the diagonal map. In this situation we identify X with the quotient $U/\mathfrak{R}$ and we will say that $\mathfrak{R} \overset{\mathfrak{s}}{\underset{\mathfrak{t}}{\rightrightarrows}} U$ is generated by γ.

Let now $\gamma : U \to X$ be a morphism of complex schemes and let

$$\mathfrak{R} \overset{\mathfrak{s}}{\underset{\mathfrak{t}}{\rightrightarrows}} U \xrightarrow{\gamma} X$$

be the presentation of X generated by γ. Let $\mathfrak{p}_1, \mathfrak{p}_2, \mathfrak{m} : \mathfrak{R} \times_U \mathfrak{R} \to \mathfrak{R}$ denote the two projections and the multiplication map respectively. Let H be an abelian group scheme over X, $\mathscr{H} \to X$ its sheaf of sections, $\mathfrak{H}$ its pullback to $\mathfrak{R}$ via $\gamma \circ \mathfrak{s} = \gamma \circ \mathfrak{t}$, and let $\pi : R \to \mathfrak{R}$ be an $\mathfrak{H}$-torsor over $\mathfrak{R}$. In order for

$$R \overset{\boldsymbol{s}}{\underset{\boldsymbol{t}}{\rightrightarrows}} U, \quad \text{with } \boldsymbol{s} := \mathfrak{s} \circ \pi, \boldsymbol{t} := \mathfrak{t} \circ \pi,$$

to be a groupoid we need a *biextension isomorphism*

$$\mathfrak{p}_1^* R \otimes \mathfrak{p}_2^* R \to \mathfrak{m}^* R$$

of torsors over $\mathfrak{R} \times_U \mathfrak{R}$, as well as a lifting

$$\boldsymbol{e} : U \to R$$

of $\mathfrak{e}$, i.e. a trivialization of $\mathfrak{e}^*R$ on U. Given these data we obtain a new presentation

$$(R \overset{s}{\underset{t}{\rightrightarrows}} U, \boldsymbol{m}, \boldsymbol{i}, \boldsymbol{e}).$$

We claim that this presentation determines an $\mathscr{H}$-gerbe $[U/R]$ on X, which we can interpret as the (stacky) quotient of U by R.

Indeed, for any open V in X, define $[U/R]'(V)$ to be the category of pairs (T, j), where T is an H-torsor over $U_{|V}$ and

$$j : \mathfrak{t}^*T \xrightarrow{\cong} \mathfrak{s}^*T \otimes R$$

is an isomorphism of $\mathfrak{H}$-torsors on $\mathfrak{R}$. As V varies, this gives a prestack $[U/R]'$ of groupoids over X. We define $[U/R]$ to be the stackification (see Remark 2.3 below) of $[U/R]'$. By construction $[U/R]'(V)$ is a torsor over the tensor category of pairs (T_0, j_0) where T_0 is an H-torsor over $U_{|V}$ and

$$j_0 : \mathfrak{t}^*T_0 \xrightarrow{\cong} \mathfrak{t}^*T_0$$

is an isomorphism of $\mathfrak{H}$-torsors on $\mathfrak{R}_{|V}$. The stackification of the latter is identified, via descent, with $B\mathscr{H}$ and so $[U/R]$ is indeed an $\mathscr{H}$-gerbe on X.

REMARK 2.3. The necessity of taking stackification in the above construction is dictated by the subtlety of the conditions required to have a 'sheaf of categories'. Let $\mathscr{X} \to X$ be a category fibered in groupoids. Recall that there are two types of sheaf-like conditions on can impose on $\mathscr{X}$:

(1): For any open $V \subset X$ and any two objects $\xi, \eta \in \mathscr{X}(V)$, the presheaf of sets (open $W \subset V$) $\mapsto \mathrm{Hom}_{\mathscr{X}(W)}(\xi_{|W}, \eta_{|W})$ is required to be a sheaf.

(2): If $V = \cup_i W_i$ is an open covering of V and we have
- $\xi_i \in \mathrm{ob}(\mathscr{X}(W_i))$;
- $\varphi_{ij} : \xi_{j|W_{ij}} \widetilde{\to} \xi_{i|W_{ij}}$ isomorphisms satisfying the cocycle condition;

then we require the existence of an object $\xi \in \mathrm{ob}(\mathscr{X}(V))$ together with isomorphisms $\psi_i : \xi_{|W_i} \widetilde{\to} \xi_i$, so that $\varphi_{ij} = \psi_i \circ \psi_j^{-1}$.

Now if $\mathscr{X}$ satisfies (1) we say that $\mathscr{X}$ is a *prestack* over X and if it satisfies (1) + (2), then we say that $\mathscr{X}(X)$ is a *stack*.

Given any prestack $\mathscr{X}$ one shows (see [**LMB00**, § 3] for details) that there is a unique (up to equivalence) stack $\mathscr{X}^a \to X$ together with a map $\mathscr{X} \to \mathscr{X}^a$ which is fully faithful and locally on X is essentially bijective. The stack $\mathscr{X}^a$ is called the stackification of $\mathscr{X}$ and is completely analogous to the sheaf one associates with a presheaf of sets.

REMARK 2.4. **(i)** We say that a groupoid $(R \underset{\boldsymbol{t}}{\overset{\boldsymbol{s}}{\rightrightarrows}} U, \boldsymbol{m}, \boldsymbol{i}, \boldsymbol{e})$ of algebraic (or analytic) spaces is *smooth* (respectively *etale*) if the structure maps $\boldsymbol{s}$ and $\boldsymbol{t}$ are smooth (respectively etale). The stacks $\mathscr{X} \to X$ over X which admit a smooth or etale groupoid presentation (lifting a presentation for X) are the stacks which are closest to schemes and on which one can do geometry in essentially the same way as on spaces. In fact if $\mathscr{X}$ is a stack which admits a smooth (respectively etale) presentation, then $\mathscr{X}$ is called an *Artin algebraic stack* (respectively *Deligne-Mumford stack*) and is the main object of study in the algebraic geometry of stacks [**LMB00**].

We consider only stacks for which the diagonal map $\mathscr{X} \to \mathscr{X} \times_X \mathscr{X}$ is affine. Note that for an $\mathscr{H}$-gerbe ${}_{\alpha}X \to X$, the condition of having an affine diagonal is equivalent to $H \to X$ being an affine group scheme. In particular ${}_{\alpha}X$ is an algebraic stack (in the sense of Artin) if and only if ${}_{\alpha}X$ has a groupoid presentation.

(ii) The case of main interest for us is when $H = \mathbb{G}_m$, so $\mathscr{H} = \mathcal{O}_X^{\times}$. In this case R is the total space of a (punctured) line bundle on $\mathfrak{R}$ and j, j_0 are isomorphisms of line bundles.

(iii) The above discussion has an obvious analogue where schemes are replaced by (algebraic or analytic) spaces or manifolds.

(iv) Not every presentation of X will lift to a presentation of a given gerbe ${}_{\alpha}X$. For example, only the trivial gerbe ${}_0X = B\mathscr{H} \to X$ can be presented by a lift of the trivial presentation $X \rightrightarrows X \to X$ generated by $\mathrm{id}_X : X \to X$. However if $\underline{\alpha} = \{\alpha_{ijk}\}$ is an $\mathscr{H}$-valued Čech cocycle w.r.t. an open covering $\{U_i\}$ of X, then the presentation of X generated by $\gamma : U = \coprod U_i \to X$ can be lifted to a presentation

$$R \underset{\boldsymbol{t}}{\overset{\boldsymbol{s}}{\rightrightarrows}} U \xrightarrow{\gamma} X$$

for an $\mathscr{H}$-gerbe ${}_{\alpha}X$, whose classifying element is the class $\alpha = [\underline{\alpha}] \in H^2(X, \mathscr{H})$. To define this presentation we take $R := \coprod_{i,j} U_{ij} \times \mathscr{H}$ with its natural projections $\boldsymbol{s}$ and $\boldsymbol{t}$ onto $U = \coprod_i U_i$. The multiplication map $\boldsymbol{m} : R \times_U R \to R$ sends a point

$$(x; a, b) \in U_{ijk} \times_X H \times_X H = (U_{ij} \times_X H) \times_{U_i} (U_{jk} \times_X H) \subset R \times_U R$$

to the point $(x; \alpha_{ijk} \cdot a \cdot b) \in U_{ik} \times_X H$. The inversion $\boldsymbol{i} : R \to R$ sends $(x, a) \in U_{ij} \times_X H$ to the point (x, a^{-1}) and the identity $\boldsymbol{e} : U \to R$ sends $x \in U_i$ to the point $(x, x; 1) \in U_{ii} \times_X H$.

Slightly more generally: the same reasoning shows that for any map of complex spaces $\gamma : U \to X$ and any $\alpha \in \check{H}^2(X, \mathscr{H})$ it follows that the presentation of X generated by γ can be lifted to a presentation for the $\mathscr{H}$-gerbe ${}_{\alpha}X$ if and only if $\gamma^*\alpha = 0 \in \check{H}^2(U, \gamma^*\mathscr{H})$.

Basic construction, continued: Assume we are given a short exact sequence of sheaves of groups

$$1 \to \mathscr{H} \to \mathscr{G} \to \mathscr{K} \to 1$$

on an algebraic (analytic) space X. Suppose that $\mathscr{H}$ is commutative and that the sheaves $\mathscr{H}$, $\mathscr{G}$ and $\mathscr{K}$ are represented by group schemes H, G and K respectively. In section 1.1 we associated to every $\mathscr{K}$ torsor $\mathscr{T} \to X$ an $\mathscr{H}$ gerbe $\delta(\mathscr{T})$ with class $\delta([\mathscr{T}]) \in H^2(X, \mathscr{H})$. In this situation the gerbe $\delta(\mathscr{T})$ comes with a natural presentation:

$$R \underset{\boldsymbol{t}}{\overset{\boldsymbol{s}}{\rightrightarrows}} U \longrightarrow \delta(\mathscr{T}).$$

Here $\gamma : U \to X$ is the scheme representing $\mathscr{T}$ and R is defined as follows. The presentation of X generated by γ has $\mathfrak{R} = U \times_X U = U \times_X K$ since U is an H-torsor. Furthermore, under the identification $\mathfrak{R} = U \times_X K$ the structure maps $\mathfrak{s}$ and $\mathfrak{t}$ become the projection on U and the action of K respectively. In other words $\mathfrak{R} \underset{\mathfrak{t}}{\overset{\mathfrak{s}}{\rightrightarrows}} U$ is the transformation groupoid for the action of K on U and $X = U/\mathfrak{R} = U/K$. To get the presentation $R \underset{\boldsymbol{t}}{\overset{\boldsymbol{s}}{\rightrightarrows}} U$ of $\delta(\mathscr{T})$ we can just take the transformation groupoid for the action of G on U (where H acts trivially), i.e. take $R := U \times_X G$ with $\boldsymbol{s}$ and $\boldsymbol{t}$ being again the projection and the action maps. Equivalently we may take $R \to \mathfrak{R}$ to be the trivial $\mathfrak{G}$-torsor and check that it satisfies the biextension and trivialization conditions. In particular we get that $\delta(\mathscr{T})$ is a quotient gerbe - it is identified as the quotient

$$\delta(\mathscr{T}) = [U/R] = [U/G]$$

of the space U by the group scheme G, where G acts with a stabilizer H at each point.

EXAMPLE 2.5. As a special case of the above we obtain the Azumaya presentation of an $\mathcal{O}_X^\times$-gerbe. Let P be a $\mathbb{P}^{n-1}$ bundle on a scheme X and let $\mathfrak{P}$ denote the corresponding sheaf of sections. The bundle P is associated to a unique $\mathbb{P}GL_n$-bundle $U \to X$ (the frame bundle of P) whose sheaf of sections $\mathscr{U}$ is naturally a torsor over $\mathbb{P}GL_n(\mathcal{O}_X)$. The image of $\mathscr{U}$ under the coboundary map for the sequence

$$1 \to \mathcal{O}_X^\times \to GL_n(\mathcal{O}_X) \to \mathbb{P}GL_n(\mathcal{O}_X) \to 1$$

is an $\mathcal{O}_X^\times$-gerbe ${}_{\mathfrak{P}}X$ on X which comes together with a *right Azumaya presentation* of ${}_{\mathfrak{P}}X$:

$$R^r \overset{s}{\underset{t}{\rightrightarrows}} U \longrightarrow {}_{\mathfrak{P}}X$$

where $R^r := U \times GL_n(\mathbb{C})$, $\boldsymbol{s} : R^r \to U$ is the projection and $\boldsymbol{t} : R^r \to U$ is the (right) action of $GL_n(\mathbb{C})$ on U. Alternatively one may consider the sheaf $\mathscr{A}_{\mathfrak{P}}$ of Azumaya algebras corresponding to $\mathfrak{P}$. The subsheaf $\mathscr{A}_{\mathfrak{P}}^\times \subset \mathscr{A}_{\mathfrak{P}}$ of invertible elements in $\mathscr{A}_{\mathfrak{P}}$ is representable by an affine group scheme $A_{\mathfrak{P}}^\times \to X$ which acts simply transitively on the left on the frame bundle $U \to X$. Using this group scheme we get a *left Azumaya presentation* of ${}_{\mathfrak{P}}X$:

$$R^l \overset{s}{\underset{t}{\rightrightarrows}} U \longrightarrow {}_{\mathfrak{P}}X$$

where $R^l := A_{\mathfrak{P}}^\times \times_X U$, $\boldsymbol{s} : R^l \to U$ is the projection and $\boldsymbol{t} : R^l \to U$ is the (left) action of $A_{\mathfrak{P}}^\times$ on U.

The same gerbe ${}_{\mathfrak{P}}X$ has yet another presentation, called the *Brauer-Severi presentation.* Here the atlas is P itself and the relations are the total space of the punctured line bundle $\mathcal{O}(1,-1)^\times$ on $P \times_X P$.

It is often useful to describe the sheaves on a gerbe as cartesian sheaves on the simplicial space generated by a presentation or equivalently as descent datum for a presentation. Concretely, given a flat presentation

$$R \overset{s}{\underset{t}{\rightrightarrows}} U \longrightarrow {}_{\alpha}X \tag{2.1}$$

of an $\mathscr{H}$-gerbe ${}_{\alpha}X$ on X we have a simple interpretation (see e.g. [**Del90**, Section 3.3] or [**LMB00**, Propositions 12.8.2 and 13.2.4] for details and proofs) for the category of sheaves on ${}_{\alpha}X$: a sheaf of $\mathcal{O}_{{}_{\alpha}X}$-modules (respectively: a line bundle, a vector bundle) on ${}_{\alpha}X$ can be identified with a pair (F, j), where F is a sheaf of $\mathcal{O}_U$-modules (respectively: a line bundle, a vector bundle, a complex of sheaves, ect.) on

U, and $j : \boldsymbol{s}^*F \widetilde{\to} \boldsymbol{t}^*F$ is an isomorphism of sheaves on R satisfying the cocycle condition[1]

$$(p_2^* j) \circ (p_1^* j) = \boldsymbol{m}^* j \tag{2.2}$$

on $R \times_{\boldsymbol{t},U,\boldsymbol{s}} R$ and the normalization $\boldsymbol{e}^*(j) = \mathrm{id}_F$ on U. Assume now that $\mathscr{H} = H(\mathcal{O}_X)$ for some complex reductive abelian group H. In this situation we can recast the above description of sheaves on ${}_\alpha X$ in terms of the presentation $\mathfrak{R} \overset{\mathfrak{s}}{\underset{\mathfrak{t}}{\rightrightarrows}} U$ generated by $\gamma : U \to X$ and the H-torsor $\pi : R \to \mathfrak{R}$. Given a sheaf of modules (F, j) on ${}_\alpha X$ we can use the map $\pi : R \to \mathfrak{R}$ to push the isomorphism j down to $\mathfrak{R}$. Decomposing according to the characters $\widehat{H}$ of H we see that $\pi_*(j)$ corresponds to a family $\{\mathfrak{j}_\chi\}_{\chi \in \widehat{H}}$ of isomorphisms, where

$$\mathfrak{j}_\chi : \mathfrak{s}^*F \otimes (\pi_* \mathcal{O}_R)_\chi \widetilde{\to} \mathfrak{t}^*F.$$

The category of sheaves of modules on ${}_\alpha X$ is therefore "graded" by the character group $\widehat{H}$. A sheaf of weight $0 \in \widehat{H}$ is just a sheaf of modules on X. In case $H = \mathbb{G}_m$ we have $\widehat{H} = \mathbb{Z}$ and the sheaves of weight n are precisely the sheaves of weight n in the sense of section 1.1. In particular, the sheaves of weight 1 are the α-twisted sheaves on X. This observation leads to a very concrete description of the weight one sheaves on a $\mathbb{G}_m$-gerbe. Starting with a presentation (2.1) of a $\mathbb{G}_m$-gerbe ${}_\alpha X$ on X write $\mathscr{L} \to \mathfrak{R} = U \times_X U$ for the line bundle associated to the $\mathbb{G}_m$-torsor $R \to \mathfrak{R}$ via the tautological character $\mathrm{id} : \mathbb{G}_m \to \mathbb{G}_m$. The groupoid condition on the presentation (2.1) gives us a biextension isomorphism $p_{12}^*\mathscr{L} \otimes p_{23}^*\mathscr{L} = p_{13}^*\mathscr{L}$ on $U \times_X U \times_X U$ and so a sheaf on ${}_\alpha X$ is the same thing as a sheaf F on U equipped with an $\mathscr{L}$-twisted descent datum on $U \times_X U$, i.e. with an isomorphism

$$p_1^*F \xrightarrow{j} p_2^*F \otimes \mathscr{L},$$

of sheaves on $U \times_X U$, satisfying the cocycle condition

$$p_{13}^* j = (p_{23}^* j \otimes \mathrm{id}_{p_{23}^*\mathscr{L}}) \circ p_{12}^* j \tag{2.3}$$

on $U \times_X U \times_X U$. Note that in writing (2.3) we had to use the biextension isomorphism for $\mathscr{L}$.

---

[1]To write this condition one uses the natural identifications

$$p_1^*\boldsymbol{s}^*F \xrightarrow{p_1^*j} p_1^*\boldsymbol{t}^*F = p_2^*\boldsymbol{s}^*F \xrightarrow{p_2^*j} p_2^*\boldsymbol{t}^*F,$$

and

$$p_1^*\boldsymbol{s}^*F = \boldsymbol{m}^*\boldsymbol{s}^*F \xrightarrow{\boldsymbol{m}^*j} \boldsymbol{m}^*\boldsymbol{t}^*F = p_2^*\boldsymbol{t}^*F,$$

provided by the groupoid axioms.

EXAMPLE 2.6. Specializing the previous discussion to the case of the Azumaya gerbe ${}_{\mathfrak{P}}X$ of example 2.5 we get a natural identification of the category $D^b_n({}_{\mathfrak{P}}X)$ with the derived category of complexes of quasi-coherent sheaves on X equipped with an action of the Azumaya algebra ${}_{\mathfrak{P}}A$ and such that the center $\mathcal{O}_X^\times$ of ${}_{\mathfrak{P}}A^\times$ acts on the cohomology sheaves with character n.

1.3. Brauer groups. Since the $\mathcal{O}^\times$ gerbes are naturally classified by elements in cohomological Brauer groups, it will be helpful to have an overview of the different variants of the Brauer group of a complex space before discussing properties of individual gerbes.

Below we are going to discuss three versions of the Brauer group of a ringed space Z: Azumaya ($Br(Z)$), geometric ($Br_{\text{geom}}(Z)$), and cohomological ($Br'(Z)$). Each of these makes sense in either the etale or the analytic topology on Z. In particular, for a complex algebraic space Z we have a diagram:

$$\begin{array}{ccccc} Br(Z) & \longrightarrow & Br(Z)_{\text{geom}} & \longrightarrow & Br'(Z) \\ \downarrow & & \downarrow & & \downarrow \\ Br_{an}(Z) & \longrightarrow & Br_{an}(Z)_{\text{geom}} & \longrightarrow & Br'_{an}(Z). \end{array}$$

When Z is *smooth*, the following facts are known:

(i) all the maps in this diagram are injective;
(ii) $Br'(Z)$ is torsion by the purity theorem from [**Gro68c**];
(iii) $\text{im}[Br'(Z) \to Br'_{an}(Z)]$ coincides (see [**Mil80**]) with the torsion subgroup of $Br'_{an}(Z)$;

Grothendieck has conjectured that the inclusion $Br(Z) \hookrightarrow Br'(Z)$ is an isomorphism for all smooth quasi-projective schemes. This may hold also for separated normal Z. The validity of the conjecture was established in the algebraic setting in [**Gab81, Hoo82, Sch01**] for arbitrary curves, for normal separated algebraic surfaces, for abelian varieties, for smooth toric varieties and for separated unions of two affine varieties. The analogous conjecture in the analytic case is virtually unexplored. The only general result to date [**HSc05**] concerns analytic K3 surfaces and asserts that every torsion class in $Br'_{an}(X)$ of an analytic K3 surface X comes from an Azumaya algerba on X.

REMARK 2.7. As a corollary of fact (i) and Grothendieck's conjecture, we get that $Br(Z) = Br(Z)_{\text{geom}}$ for a smooth Z. This corollary is known to hold [**EHKV01**] in many cases in which the Grothendieck conjecture is still unknown. In fact, for a normal Noetherian scheme, the result of [**EHKV01**, Theorem 3.6] characterizes the image of $Br(Z)$ in $Br(Z)_{\text{geom}}$ as the algebraic-geometric gerbes for which one can find a flat presentation

$$\left(R \overset{s}{\underset{t}{\rightrightarrows}} U, \boldsymbol{m}, \boldsymbol{i}, \boldsymbol{e}, \boldsymbol{\gamma} \right)$$

with a projective structure map $\boldsymbol{\gamma} : U \to Z$, or equivalently as the classes of $\mathbb{G}_m$ gerbes of quotient type (i.e. a quotient of an algebraic space by an affine algebraic group). In general this characterization seems to be optimal since there are examples of quotient gerbes on non-separated surfaces whose isomorphism class is not represented by an Azumaya algebra, and examples of infinite order elements in $Br'(Z)$ for a normal separated Z which are represented by algebraic-geometric gerbes but are not quotient gerbes [**EHKV01**, Examples 2.21 and 3.12].

If Z is a complex scheme, then the *Azumaya Brauer group* $Br(Z)$, is defined [**Gro68a**] as the group of Morita equivalence classes of sheaves of Azumaya algebras on Z. Recall [**Gro68a**] that an Azumaya algebra on Z is a coherent sheaf of $\mathcal{O}_Z$-algebras which locally in the etale topology on Z is isomorphic to the endomorphisms algebra of an algebraic vector bundle on Z. Two Azumaya algebras $\mathcal{A}$ and $\mathcal{B}$ are called Morita equivalent if etale locally on Z we can find vector bundles E and F so that the sheaves of algebras $\mathcal{A} \otimes \mathcal{E}nd(E)$ and $\mathcal{B} \otimes \mathcal{E}nd(F)$ are isomorphic. Morita equivalence classes of Azumaya algebras form a commutative group under the operation of tensoring over $\mathcal{O}_Z$; the inverse is given by the opposite algebra.

The Skolem-Noether theorem [**Mil80**, Proposition 2.3] implies that the Azumaya algebras of rank n^2 are classified by elements in $H^1_{et}(Z, \mathbb{P}GL(n))$. The short exact sequence of groups schemes over Z:

$$1 \to \mathbb{G}_m \to GL(n) \to \mathbb{P}GL(n) \to 1,$$

gives rise to a coboundary map

$$H^1_{et}(Z, \mathbb{P}GL(n)) \to H^2_{et}(Z, \mathbb{G}_m). \tag{2.4}$$

The image of $a \in H^1_{et}(Z, \mathbb{P}GL(n))$ under this coboundary map is an n-torsion class in $H^2_{et}(Z, \mathbb{G}_m)$ which is the obstruction to representing a by the endomorphism algebra of a rank n vector bundle. In particular the map (2.4) induces a homomorphism

$$Br(Z) \to H^2_{et}(Z, \mathbb{G}_m)_{\text{tor}} \subset H^2_{et}(Z, \mathbb{G}_m). \tag{2.5}$$

When Z is smooth, the homomorphism (2.5) is known to be injective [**Mil80**, Theorem IV.2.5]. This suggests that the Brauer classes are intimately related to elements in $H^2_{et}(Z, \mathbb{G}_m)$ and so one defines the *algebraic cohomological Brauer group*:

$$Br'(Z) := H^2_{et}(Z, \mathbb{G}_m).$$

Recall that by fact (ii) the group $H^2_{et}(Z, \mathbb{G}_m)$ is purely torsion. As explained in section 1.2, Azumaya algebras give rise to groupoid presentations of $\mathbb{G}_m$-gerbes on Z. In other words, for a smooth Z the inclusion (2.5) can be refined to a sequence of inclusions:

$$Br(Z) \hookrightarrow Br(Z)_{\mathrm{geom}} \hookrightarrow Br'(Z) = H^2_{et}(Z, \mathbb{G}_m) = H^2_{et}(Z, \mathbb{G}_m)_{\mathrm{tor}},$$

where $Br(Z)_{\mathrm{geom}}$ denotes the group of equivalence classes of algebraic-geometric $\mathbb{G}_m$-gerbes on Z. Recall that a $\mathbb{G}_m$-gerbe is *algebraic geometric* if it is an algebraic stack in the sense of Artin, i.e. if it admits a flat (equivalently, a smooth) groupoid presentation [**Art74**].

By analogy we define the *analytic Azumaya Brauer group* $Br_{an}(Z)$, and the *analytic geometric Brauer group* $Br_{an}(Z)_{\mathrm{geom}}$ of an analytic space Z, as the groups of Morita equivalent classes of analytic Azumaya algebras on Z and of isomorphism classes of analytic geometric $\mathcal{O}^{\times}_{Z_{an}}$-gerbes respectively. The isomorphism type of an $\mathcal{O}^{\times}_{Z_{an}}$-gerbe is determined by a class in the *analytic cohomological Brauer group*:

$$Br'_{an}(Z) := H^2_{an}(Z, \mathcal{O}^{\times}_Z).$$

In other words, the classifying map

$$Br_{an}(Z)_{\mathrm{geom}} \hookrightarrow Br'_{an}(Z)$$

is injective.

The analytic cohomological Brauer group can be studied via the exponential sequence:

$$0 \to \mathbb{Z} \to \mathcal{O} \stackrel{\exp}{\to} \mathcal{O}^{\times} \to 1.$$

The corresponding cohomology sequence gives

$$0 \longrightarrow H^2_{an}(Z, \mathcal{O}_Z)/H^2(Z, \mathbb{Z}) \longrightarrow Br'_{an}(Z) \longrightarrow \ker\left[H^3(Z, \mathbb{Z}) H^3_{an}(Z, \mathcal{O}_Z)\right] \longrightarrow 0$$

This of course is analogous to the usual description of the Picard group:

$$0 \longrightarrow H^1_{an}(Z, \mathcal{O}_Z)/H^1(Z, \mathbb{Z}) \longrightarrow \operatorname{Pic}(Z) \longrightarrow \underbrace{\ker\left[H^2(Z, \mathbb{Z}) \to H^2_{an}(Z, \mathcal{O}_Z)\right]}_{\| \ \boxed{\text{when } Z \text{ is compact and Kähler}}\atop H^{1,1}_{\mathbb{Z}}(Z)} \longrightarrow 0$$

is a discrete subgroup of maximal rank. Hence, we can identify the connected component of $\operatorname{Pic}(Z)$ with the quotient of its tangent space $H^1_{an}(Z, \mathcal{O}_Z)$ by $H^1(Z, \mathbb{Z})$. In the case of $Br'_{an}(Z)$, there is still a 'tangent space': $H^2_{an}(Z, \mathcal{O}_Z)$, but it is divided by the typically non-discrete subgroup

$$\operatorname{im}[H^2(Z, \mathbb{Z}) \to H^2_{an}(Z, \mathcal{O}_Z)] \subset H^2_{an}(Z, \mathcal{O}_Z),$$

and so there is no good (=separated) topology on $Br'_{an}(Z)$.

In the special case when Z is a $K3$ surface, we get that $Br_{an}(Z)_{\text{geom}} = Br'_{an}(Z)$ is the quotient of the one dimensional vector space $H^2_{an}(Z, \mathcal{O}_Z)$ by the lattice dual to the transcendental lattice of Z, i.e. by $H^2(Z, \mathbb{Z})/H^{1,1}_{\mathbb{Z}}(Z)$. Notice that for a very general analytic $K3$ this lattice has rank 22 and for a very general algebraic $K3$ it has rank 21.

More precisely, one defines the transcendental lattice $\boldsymbol{T}_Z$ of a $K3$ surface Z by the short exact sequence:

$$0 \to \boldsymbol{T}_Z \to H^2(Z, \mathbb{Z}) \to H^{1,1}_{\mathbb{Z}}(Z)^\vee \to 0.$$

In other words, $\boldsymbol{T}_Z$ is the sublattice of $H^2(Z, \mathbb{Z})$ consisting of classes perpendicular to all integral $(1,1)$ classes on Z. The dual sequence reads:

$$0 \to H^{1,1}_{\mathbb{Z}}(Z) \to H^2(Z, \mathbb{Z}) \to \boldsymbol{T}_Z^\vee \to 0,$$

and we have a natural map

$$\begin{array}{ccc} \operatorname{Hom}_{\mathbb{Z}}(\boldsymbol{T}_Z, \mathbb{R}) & & H^2_{an}(Z, \mathcal{O}_Z) \\ \| & & \| \\ H^2(Z, \mathbb{R})/(H^{1,1}_{\mathbb{Z}}(Z) \otimes \mathbb{R}) & \twoheadrightarrow & H^2(Z, \mathbb{R})/H^{1,1}_{\mathbb{R}}(Z) \end{array}$$

This leads to the following commutative diagram with exact rows and columns:

$$
\begin{array}{ccccccc}
 & & & & 0 & & 0 \\
 & & & & \downarrow & & \downarrow \\
 & & & & \mathrm{Hom}_{\mathbb{Z}}(\boldsymbol{T}_Z, \mathbb{Z}) & \xrightarrow{\cong} & \boldsymbol{T}_Z^{\vee} \\
 & & & & \downarrow & & \downarrow \\
0 \to & H^{1,1}_{\mathbb{R}}(Z)/(H^{1,1}_{\mathbb{Z}}(Z) \otimes \mathbb{R}) & \longrightarrow & & \mathrm{Hom}_{\mathbb{Z}}(\boldsymbol{T}_Z, \mathbb{R}) & \longrightarrow & H^2_{an}(Z, \mathcal{O}_Z) \to 0 \\
 & \| & & & \downarrow & & \downarrow \\
0 \to & H^{1,1}_{\mathbb{R}}(Z)/(H^{1,1}_{\mathbb{Z}}(Z) \otimes \mathbb{R}) & \to & & \mathrm{Hom}_{\mathbb{Z}}(\boldsymbol{T}_Z, \mathbb{R}/\mathbb{Z}) & \longrightarrow & Br'_{an}(Z) \longrightarrow 0 \\
 & & & & \downarrow & & \downarrow \\
 & & & & 0 & & 0
\end{array}
$$

The bottom row explicates $Br_{an}(Z) = Br'_{an}(Z)$ as the quotient of the real torus $\mathrm{Hom}_{\mathbb{Z}}(\boldsymbol{T}_Z, \mathbb{R}/\mathbb{Z})$ by the vector space $H^{1,1}_{\mathbb{R}}(Z)/(H^{1,1}_{\mathbb{Z}}(Z) \otimes \mathbb{R})$, embedded in it as a (usually dense) subgroup. Note that this vector space does not contain any torsion points of the torus. Equivalently the restricted map

$$\mathrm{Hom}_{\mathbb{Z}}(\boldsymbol{T}_Z, \mathbb{Q}/\mathbb{Z}) \widetilde{\to} Br_{an}(Z)_{\mathrm{tor}}$$

is an isomorphism. When Z happens to be an algebraic $K3$ surface we have a natural identification $Br(Z) = Br_{an}(Z)_{\mathrm{torsion}}$ and so we recover the standard interpretation of elements of the algebraic Brauer group of Z as a homomorphism from the transcendental lattice of Z to $\mathbb{Q}/\mathbb{Z}$ (see e.g. [**Căl00**, Lemma 5.4.1] or [**Căl02a**]).

2. Tate-Shafarevich groups and genus one fibrations

In this section we review some basic facts about twisted forms of a given elliptic fibration over an analytic space B. For more details the reader is referred to the excellent references [**DG94**] and [**Nak02**]. First we recall some terminology and set up the notation.

For us a *genus one fibration* will always mean a holomorphic map $\pi : X \to B$ between normal analytic varieties whose generic fiber is a smooth curve of genus one. We define an *elliptic fibration* to be a genus one fibration equipped with a holomorphic section $\sigma : B \to X$ of π. Note that this is slightly more restrictive than the conventional notion of an elliptic fibration used in say [**DG94**], [**Nak02**], where only the existence of a meromorphic section of π is required. A genus one fibration will be called (relatively) minimal if X has at most terminal singularities and if the canonical class K_X is π-nef.

Let now X and B be normal analytic varieties and let

$$X \underset{\sigma}{\overset{\pi}{\rightleftarrows}} B$$

be an elliptic fibration on X. Let $D \subset B$ denote the discriminant divisor of π and let $B^o := B - D$, $B^{oo} := B - \mathrm{Sing}(D)$. The corresponding inclusions are denoted by $\imath : D \hookrightarrow B$, $\jmath^o : B^o \hookrightarrow B$ and $\jmath^{oo} : B^{oo} \hookrightarrow B$. We also put $X^o := X \times_B B^o$, $X^{oo} := X \times_B B^{oo}$ and $\pi^o := \pi_{|X^o}$, $\pi^{oo} := \pi_{|X^{oo}}$.

Sometimes we may need to require the additional genericity assumption that X is smooth and that $\pi : X \to B$ is Weierstrass.

REMARK 2.8. **(i)** When X is a surface, the genericity assumption implies in particular that all the singular fibers of π are of Kodaira types I_1 or II, i.e. they are nodes and cusps.

(ii) In this paper we will always deal with a situation in which X is smooth and either π is smooth or X is a surface and π has at worst I_1 fibers. We have included in the present discussion the more general case of an arbitrary Weierstrass π with a smooth total space, because of the potential applications of our duality construction to genus one fibered Calabi-Yau manifolds of arbitrary dimension. This however goes beyond the scope of the present work and will be the subject of a future paper.

Let $X^\sharp \subset X$ denote the regular locus of π, viewed as an abelian group scheme over B. Denote by $\mathscr{X}_{an}$ the corresponding sheaf of abelian groups in the analytic topology on B. When B and X happen to underly complex algebraic varieties we will write $\mathscr{X}$ for the etale sheaf of sections of $X^\sharp \to B$.

The *analytic Weil-Châtelet group* $WC_{an}(X)$ of X is defined (see e.g. [**Nak02**]) as the group of bimeromorphism classes of analytic genus one fibrations $Y \to B$ such that:

- $Y \times_B B^o \to B^o$ is bimeromorphic to a smooth genus one fibration;
- The relative Jacobian fibration $\mathrm{Pic}^0(Y/B)$ is bimeromorphic to $X^\sharp$ (and hence to X). Note that this definition makes sense since for a suitably chosen dense open subset $U \subset B$

the (sheafification of the) presheaf $\mathscr{P}ic^0(Y/U)$ of relative Picard groups along the fibers of $Y \times_B U \to U$ is representable by an analytic space [**Nak02**].

The *analytic Tate-Shafarevich group* $Ш_{an}(X)$ of X is the subgroup of $WC_{an}(X)$ consisting of elements $\alpha \in WC_{an}(X)$ such that for any representative $Y \to B$ of α and any point $b \in B$ one can find an analytic neighborhood $b \in U \subset B$ so that $Y \times_B U \to U$ has a meromorphic section. This implies that $Y \to B$ has no multiple fibers in codimension one.

The group $Ш_{an}(X)$ can be described cohomologically [**Nak02**] as follows. Assume that X^{oo} is a smooth space. Then by [**Nak02**, Proposition 5.5.1], the natural classifying map

$$Ш_{an}(X) \to H^1_{an}(B, j^{oo}_* j^{oo*}\mathscr{X}_{an}). \tag{2.6}$$

is injective. Furthermore if $B^{oo} = B$, or if π is Weierstrass with a smooth total space, then the map (2.6) is an isomorphism [**Nak02**, Proposition 5.5.1]. In addition one knows (see e.g. [**Nak02**, Theorem 5.4.9]) that under the same assumptions, the sheaf $j^{oo}_* j^{oo*}\mathscr{X}_{an}$ fits in a short exact sequence

$$0 \to \mathscr{X}_{an} \to j^{oo}_* j^{oo*}\mathscr{X}_{an} \to (R^2\pi_*\mathbb{Z}_X/\imath_*\imath^! R^1\pi_*\mathcal{O}_X^\times)_{\text{torsion}} \to 0.$$

Since by definition the sheaf $(R^2\pi_*\mathbb{Z}_X/\imath_*\imath^! R^1\pi_*\mathcal{O}_X^\times)_{\text{torsion}}$ is supported on the multiple fiber sublocus of D, it follows that in the absence of multiple fibers, i.e. under our definition of an elliptic fibration we have an isomorphism:

$$Ш_{an}(X) \cong H^1_{an}(B, \mathscr{X}_{an}). \tag{2.7}$$

In the remainder of this paper we will always assume tacitly that the isomorphism (2.7) holds, in fact we will assume that either π is smooth or that X is a surface.

Because of this cohomological interpretation we can view the elements in $Ш_{an}(X)$ simply as $\mathscr{X}_{an}$-torsors. This definition of $Ш_{an}(X)$ is consistent with the usual definition of the algebraic Tate-Shafarevich group [**Gro68a, Gro68b, Gro68c**] and [**DG94**]. When X is a surface we can also interpret (by compactifying a genus one fibration and then choosing a Weierstrass smooth model over B) the elements in $Ш_{an}(X)$ as smooth analytic surfaces equipped with a genus one fibration over B.

The *algebraic Weil-Châtelet and Tate-Shafarevich groups* $WC(X)$ and $Ш(X)$ are defined in a similar manner [**Gro68a, Gro68b, Gro68c**] and [**DG94**] with the etale topology replacing the analytic one. Furthermore, the analysis carried out in [**DG94**, Section 1] implies, that under the assumption that X and B are both smooth and that π has a regular section, the algebraic Tate-Shafarevich group can be interpreted cohomologically as

$$Ш(X) = H^1_{et}(B, \mathscr{X}),$$

i.e. the elements in $Ш(X)$ can be viewed as algebraic spaces $Y \to B$ which are $\mathscr{X}$-torsors.

Given an element $\alpha \in Ш_{an}(X)$ (or $\alpha \in Ш(X)$) we denote by $X^\sharp_\alpha$ the analytic (or algebraic) space representing the torsor α and by $\pi^\sharp_\alpha : X^\sharp_\alpha \to B$ the corresponding projection. Following [**DG94**] we say that a morphism of analytic (algebraic) spaces $Y \to B$ is a *good model* for α if $Y \to B$ is bimeromorphic to $X^\sharp_\alpha \to B$, Y is smooth and the map $Y \to B$ is proper and flat.

REMARK 2.9. Note that when π is smooth $X^\sharp_\alpha$ is itself a good model for α and when X is an arbitrary smooth surface we always have a preferred good model for α, namely the relatively minimal model of a compactification of $X^\sharp_\alpha$. When X is of dimension three the good models of elements in $Ш(X)$ have been analyzed in detail, see e.g. [**Mir83, Gra91, DG94**]. In this case the good model exists (possibly after blowing up B at finitely many points) but is not unique. However all good models of a given α are related by flops and in particular have equivalent derived categories of coherent sheaves (see e.g. [**BO95, Bri02, Kaw02**]).

In the cases when $\pi : X \to B$ is smooth or X is a surface we put

$$\pi_\alpha : X_\alpha \to B$$

for the canonical good model of α. In particular, if $\pi : X \to B \cong \mathbb{P}^1$ is an elliptic $K3$ surface we have that X_α is a well defined analytic (respectively algebraic) $K3$ surface for any element $\alpha \in Ш_{an}(X)$ (respectively $\alpha \in Ш(X)$).

The meromorphic action of the analytic group space $X^\sharp \to B$ on X_α induces a natural meromorphic action map

$$a_\alpha : X \times_B X_\alpha \dashrightarrow X_\alpha.$$

Furthermore, given a positive integer n we can consider the sheaf of groups $\mathscr{X}_{an}[n] \to B$ consisting of the n-torsion points in $\mathscr{X}_{an}$. The sheaf $\mathscr{X}_{an}[n]$ is represented by a group space $X^\sharp[n]$ which is quasi-finite over B. We will write $X[n]$ for the closure of $X^\sharp[n]$ in X and by an abuse of notation we will denote the meromorphic map

$$X[n] \times_B X_\alpha \dashrightarrow X_\alpha$$

again by a_α.

Since $X^\sharp[n]$ is finite over a dense open set in B we can form the quotient $X_\alpha/X^\sharp[n]$ which as an analytic space is well defined up to a bimeromorphism which respects the genus one fibration. Moreover $X_\alpha/X^\sharp[n]$ is naturally a $\mathscr{X}_{an}$-torsor at the general point and so represents an element in $Ш_{an}(X)$. It is not hard to calculate this element in terms of α and n only. In fact it is clear that $X_\alpha/X^\sharp[n]$ is tautologically the same as the quotient

$$X_\alpha^{\times_B n}/K,$$

which by definition represents the element $n\alpha \in Ш_{an}(X)$.

Here

$$K = \ker[(X^\sharp)^{\times_B n} \xrightarrow{\text{mult}} X^\sharp]$$

is the kernel of the natural product map corresponding the group law on $X^\sharp$, and the action of K is induced from the component-wise action of $(X^\sharp)_B^{\times n}$ on $(X_\alpha)_B^{\times n}$.

In particular we have a bimeromorphism $X_\alpha/X^\sharp[n] \widetilde{\dashrightarrow} X_{n\alpha}$ which is unique up to an auto-bimeromorphism of $X_{n\alpha}$, compatible with the genus one fibration. However, as one can see from the proof of [**Nak02**, Lemma 5.3.3], if we assume that π is relatively minimal with a smooth total space, then all such auto-bimeromorphisms are holomorphic and are translations by sections in $\mathscr{X}_{an}$. So, under this the genericity assumption, we will have a bimeromorphic identification $X_{n\alpha} = X_\alpha/X^\sharp[n]$ and hence a well defined meromorphic map

$$q_\alpha^n : X_\alpha \dashrightarrow X_{n\alpha}.$$

If in addition we assume that the fibration $\pi : X \to B$ has a trivial Mordel-Weil group, then the meromorphic map q_α^n is canonical and does not depend on any choices.

The *index* of an element $\alpha \in Ш_{an}(X)$ is defined to be the minimal degree of a global multisection of π_α. We will denote the index by $\text{ind}(\alpha)$.

Assume now that B and X are quasi-projective. Since the element $0 \in Ш_{an}(X)$ is represented by the algebraic elliptic fibration $\pi : X \to B$, it follows that for each α of finite index the space X_α admits a dominant meromorphic map

$$q_\alpha^{\text{ind}(\alpha)} : X_\alpha \dashrightarrow X$$

to the algebraic variety X. In fact in [**Nak02**, Proposition 5.5.4] Nakayama proves that such a X_α is bimeromorphic to an algebraic variety and so must be an algebraic space. Furthermore in the case of surfaces Kodaira shows [**Kod63**] that X_α is quasi-projective if and only if α is torsion in $\text{Ш}_{an}(X)$.

3. Complementary fibrations

Let X be smooth and let

$$X \underset{\sigma}{\overset{\pi}{\rightleftarrows}} B$$

be a relatively minimal elliptic fibration. Consider an element $\alpha \in \text{Ш}_{an}(X)$ and a good representative $\pi_\alpha : X_\alpha \to B$ for α. Our goal in this section is to describe the cohomological Brauer group $Br'_{an}(X_\alpha)$ in terms of the Tate-Shafarevich group $\text{Ш}_{an}(X)$. For this we need to analyze the relationship between the sheaf $\mathscr{X}_{an}$ and the relative Picard sheaf of π_α.

If all the fibers of π are integral, then $\mathscr{P}ic(X_\alpha/B)$ is representable and we have a short exact sequence of abelian sheaves in the analytic topology:

$$0 \to \mathscr{X} \to \mathscr{P}ic(X_\alpha/B) \xrightarrow{\deg_\alpha} \mathbb{Z} \to 0, \tag{2.8}$$

where $\deg_\alpha$ is the map assigning to each $L \in \operatorname{Pic}(\pi_\alpha^{-1}(U))/\pi_\alpha^* \operatorname{Pic}(U)$ its degree along a smooth fiber.

REMARK 2.10. If we want to allow non-integral fibers for π, then $\mathscr{P}ic(X_\alpha/B)$ becomes non-representable, but it has a maximal representable quotient $\mathscr{Q}_\alpha$ as shown in e.g. [**Ray70**] and [**DG94**] in the algebraic case and [**Nak02**] in the analytic case. The sheaf of groups $\mathscr{Q}_\alpha$ is defined as:

$$\mathscr{Q}_\alpha := \mathscr{P}ic(X_\alpha/B)/\mathscr{E}_\alpha,$$

where $\mathscr{E}_\alpha \subset \mathscr{P}ic(X_\alpha/B)$ is a subsheaf generated by local components of the preimage $\pi_\alpha^{-1}(D)$ of the discriminant $D \subset B$ (see [**DG94**, Proposition 1.13] for the precise statement). Note that when all fibers of π are integral we have $\mathscr{E}_\alpha = 0$.

In this generality, the short exact sequence (2.8) is replaced by a commutative diagram with exact rows and columns:

$$
\begin{array}{ccccccccc}
 & & & & 0 & & 0 & & \\
 & & & & \uparrow & & \uparrow & & \\
 & & & & \mathbb{Z} & = & \mathbb{Z} & & \\
 & & & & \uparrow{\scriptstyle \deg_\alpha} & & \uparrow & & \\
0 & \longrightarrow & \mathscr{X}_{an} & \longrightarrow & \mathscr{Q}_\alpha & \longrightarrow & R^2\pi_{\alpha *}\mathbb{Z}/\mathscr{E}_\alpha & \longrightarrow & 0 \\
 & & \| & & \uparrow & & \uparrow & & \\
0 & \longrightarrow & \mathscr{X} & \longrightarrow & \ker(\deg_\alpha) & \longrightarrow & (R^2\pi_{\alpha *}\mathbb{Z}/\mathscr{E}_\alpha)_{\text{torsion}} & \longrightarrow & 0 \\
 & & & & \uparrow & & \uparrow & & \\
 & & & & 0 & & 0 & &
\end{array}
$$

Note also that $(R^2\pi_{\alpha *}\mathbb{Z}/\mathscr{E}_\alpha)_{\text{torsion}}$ is supported on D and that its fibers at smooth points of a component of D parameterizing Kodaira fibers of type I_n are isomorphic to $\mathbb{Z}/n$.

Fix now an element $\alpha \in Ш_{an}(X)$. Under some mild assumptions on α we will construct a natural map $T_\alpha : Ш_{an}(X) \to Br'_{an}(X_\alpha)$, which will allow us to compare the Tate-Shafarevich and Brauer groups. The existence of T_α is established in the following lemma.

LEMMA 2.11. *Assume that X is smooth, π has integral fibers and $Br'_{an}(B) = 0$. Assume further that*

$$\ker\left(H^3_{an}(B, \mathscr{O}_B^\times) \xrightarrow{\pi_\alpha^*} H^3_{an}(X_\alpha, \mathscr{O}_{X_\alpha}^\times)\right) = 0 \tag{2.9}$$

Then there is a canonical homomorphism T_α which fits in an exact sequence of abelian groups:

$$0 \to \mathbb{Z}/\operatorname{ind}(\alpha) \to Ш_{an}(X) \xrightarrow{T_\alpha} Br'_{an}(X_\alpha) \to H^1(B, \mathbb{Z}).$$

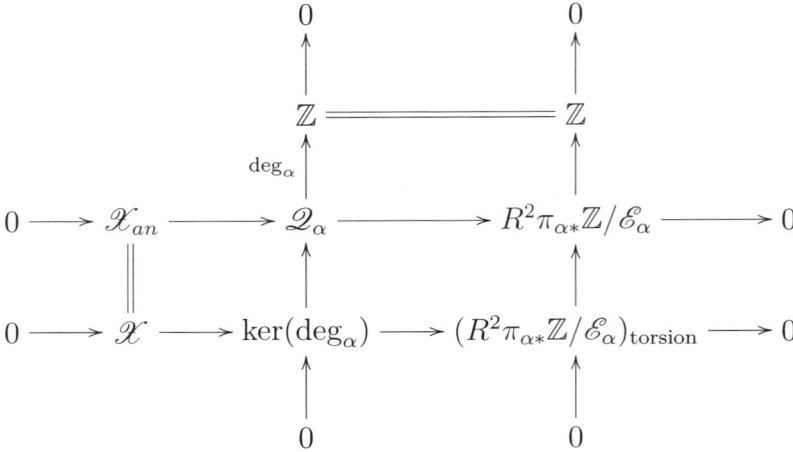

Proof. The long exact sequence of (2.8) gives

$$\begin{array}{c} \mathbb{Z}/\operatorname{ind}(\alpha) \\ \| \\ 0 \longrightarrow \overbrace{H^0(B,\mathbb{Z})/H^0_{an}(B,\mathscr{P}ic(X_\alpha/B))} \end{array}$$

$$H^1_{an}(B,\mathscr{X}) \longrightarrow H^1_{an}(B,\mathscr{P}ic(X_\alpha/B)) \longrightarrow H^1(B,\mathbb{Z})$$

$$\begin{array}{c} \| \\ Ш_{an}(X) \end{array}$$

So it suffices to find an identification

$$H^1_{an}(B,\mathscr{P}ic(X_\alpha/B)) \cong Br'_{an}(X_\alpha). \tag{2.10}$$

Consider now the Leray spectral sequence for $\pi_\alpha : X_\alpha \to B$ and the sheaf $\mathcal{O}^\times_{X_\alpha}$, which has only two non-zero rows, so it also becomes a long exact sequence:

$$Br'_{an}(B) \longrightarrow Br'_{an}(X_\alpha)$$

$$\longrightarrow H^1_{an}(B,\mathscr{P}ic(X_\alpha/B)) \to \ker\left(H^3_{an}(B,\mathcal{O}^\times_B) \xrightarrow{\pi^*_\alpha} H^3_{an}(X_\alpha,\mathcal{O}^\times_{X_\alpha})\right).$$

The assumption $\ker\left(H^3_{an}(B,\mathcal{O}^\times_B) \xrightarrow{\pi^*_\alpha} H^3_{an}(X_\alpha,\mathcal{O}^\times_{X_\alpha})\right) = 0$ thus immediately implies the identification (2.10) and so the lemma is proven. □

The lemma has the following immediate corollary:

COROLLARY 2.12. *Assume that X is a smooth projective surface and π has integral fibers. Then the map*

$$T_\alpha : Ш_{an}(X) \to Br'_{an}(X_\alpha)$$

exists for all $\alpha \in Ш_{an}(X)$.

Proof. The existence of T_α is an immediate consequence of Lemma 2.11 since in this case B is a smooth curve and so $H^2_{an}(B,\mathcal{O}^\times_B) = H^3_{an}(B,\mathcal{O}^\times_B) = 0$ for dimension reasons. □

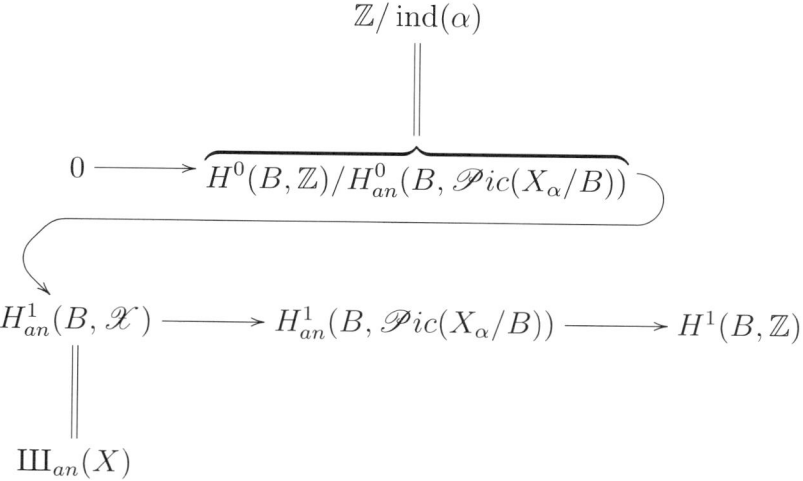

If the vanishing assumption (2.9) does not hold, we can still construct a variant of the map T_α which is defined only on a part of the group $Ш_{an}(X)$:

LEMMA 2.13. *Assume that X is smooth, π has integral fibers and $Br'_{an}(B) = 0$. Then:*

(i) *if α is m-torsion in $Ш_{an}(X)$, then there is a group homomorphism (compatible with T_α when the latter exists)*
$$mШ_{an}(X) \to Br'_{an}(X_\alpha),$$
from the subgroup $mШ_{an}(X) \subset Ш_{an}(X)$ of m-divisible elements in $Ш_{an}(X)$ to the cohomological Brauer group of X_α;

(ii) *if α is m-divisible in $Ш_{an}(X)$, then there is a group homomorphism (compatible with T_α when the latter exists)*
$$Ш_{an}(X)[m] \to Br'_{an}(X_\alpha)$$
from the subgroup $Ш_{an}(X)[m] \subset Ш_{an}(X)$ of m-torsion elements in $Ш_{an}(X)$ to the cohomological Brauer group of X_α;

Proof. For any given $\alpha \in Ш_{an}(X)$ we have a composition map

$$\begin{array}{ccc} Ш_{an}(X) & \xrightarrow{\quad d_\alpha \quad} & H^3_{an}(B, \mathcal{O}^\times_B) \\ & \searrow \qquad \nearrow & \\ & H^1_{an}(B, \mathscr{P}ic(X_\alpha/B)) & \end{array}$$

The assignment $\alpha \mapsto d_\alpha$ gives rise to a group homomorphism

$$d : Ш_{an}(X) \to \mathrm{Hom}_{\mathbb{Z}}(Ш_{an}(X), H^3_{an}(B, \mathcal{O}^\times_B)). \tag{2.11}$$

In particular, if α is m-torsion, then $d_\alpha(m \cdot \xi) = d_{m\cdot\alpha}(\xi) = d_0(\xi) = 0 \in H^3_{an}(B, \mathcal{O}^\times_B)$ and so the image of $mШ_{an}(X)$ in $H^1_{an}(B, \mathscr{P}ic(X_\alpha/B))$ must be contained in $Br'_{an}(X_\alpha)$. This proves *(i)*.

Similarly, if $\alpha = m\cdot\varphi$ is m-divisible, then for any ξ we have $d_\alpha(\xi) = d_{m\cdot\varphi}(\xi) = d_\varphi(m\cdot\xi)$ and so d_α vanishes identically on $Ш_{an}(X)[m]$. Thus the image of $Ш_{an}(X)[m]$ in $H^1_{an}(B, \mathscr{P}ic(X_\alpha/B))$ must be contained in $Br'_{an}(X_\alpha)$ which completes the proof of *(ii)* and the lemma. □

We will denote the maps in items *(i)* and *(ii)* of Lemma 2.13 again by T_α. Since by construction these maps are compatible with the map T_α from Lemma 2.11, whenever the latter exists, this abuse of notation can not lead to any confusion.

Let us examine in more detail the map

$$d : H^1_{an}(B, \mathscr{X}) \to \mathrm{Hom}_{\mathbb{Z}}(H^1_{an}(B, \mathscr{X}), H^3_{an}(B, \mathcal{O}_B^{\times}))$$

given in (2.11). This map can be rewritten as a bilinear pairing

$$\langle \bullet, \bullet \rangle : H^1_{an}(B, \mathscr{X}) \otimes_{\mathbb{Z}} H^1_{an}(B, \mathscr{X}) \to H^3_{an}(B, \mathcal{O}_B^{\times}).$$

The proof of Lemma 2.13 shows that for every $\alpha \in Ш_{an}(X)$ we have a well defined homomorphism

$$T_\alpha : \alpha^\perp \to Br'_{an}(X_\alpha),$$

where $\alpha^\perp \subset Ш_{an}(X)$ is the orthogonal complement of α with respect to $\langle \bullet, \bullet \rangle$.

DEFINITION 2.14. Two genus one fibrations $\alpha, \beta \in Ш_{an}(X)$ will be called *complementary* if $\langle \alpha, \beta \rangle = 0$. We will call α and β *m-compatible* if one of them is m-divisible and the other one is m-torsion.

Note that using the pairing $\langle \bullet, \bullet \rangle$, Lemma 2.13 follows from the obvious observation that every m-compatible pair α, β is complementary.

For future reference we spell out the special case when $\alpha = 0$:

COROLLARY 2.15. *Assume that X is smooth, the fibers of π are integral, and $Br'_{an}(B) = 0$. Then we have an isomorphism $H^1_{an}(B, \mathscr{P}ic(X/B)) \cong Br'_{an}(X)$ and we have an exact sequence of abelian groups*

$$0 \to Ш_{an}(X) \xrightarrow{T_0} Br'_{an}(X) \to H^1(B, \mathbb{Z}).$$

Proof. Since $\sigma : B \to X$ is a section of π it follows that the composition

$$H^i_{an}(B, \mathcal{O}_B^{\times}) \xrightarrow{\pi^*} H^i_{an}(X, \mathcal{O}_X^{\times}) \xrightarrow{\sigma^*} H^i_{an}(B, \mathcal{O}_B^{\times})$$

is the identity. Thus $\ker\left(H^i_{an}(B, \mathcal{O}_B^{\times}) \xrightarrow{\pi^*} H^i_{an}(X, \mathcal{O}_X^{\times})\right) = 0$ and so $H^1_{an}(B, \mathscr{P}ic(X/B)) \cong Br'_{an}(X)$. Combined with the fact that $\mathrm{ind}(0) = 1$ this gives the short exact sequence of groups above. The corollary is proven. □

Our pairing $\langle \bullet, \bullet \rangle$ can be explicitly described as follows. Every element $\alpha \in Ш_{an}(X) = H^1_{an}(B, \mathscr{X})$ has two different incarnations:

- α can be interpreted as a group extension of $\mathbb{Z}_B$ by $\mathscr{X}$. Concretely this is just the sheaf of groups $\mathscr{P}ic(X_\alpha/B)$ as it fits in the extension (2.8) viewed as an element $e(\alpha)$ in $\mathrm{Ext}^1_{\mathbb{Z}_B}(\mathbb{Z}_B, \mathscr{X})$.

- α can be interpreted as an extension of $\mathscr{X}$ by $\mathcal{O}_B^\times[1]$. Concretely this is the amplitude one object ${}_\alpha\mathscr{X}$ in the derived category of abelian sheaves on B which is the pullback of the extension class of
$$1 \to \mathcal{O}_B^\times[1] \to R\pi_{\alpha *}\mathcal{O}_{X_\alpha}^\times[1] \to R^1\pi_{\alpha *}\mathcal{O}_{X_\alpha}^\times \to 1$$
via the natural inclusion $\mathscr{X} \to \mathscr{P}ic(X_\alpha/B) = R^1\pi_{\alpha *}\mathcal{O}_{X_\alpha}^\times$. Alternatively, ${}_\alpha\mathscr{X}$ can be thought of as a sheaf of commutative group stacks on B which is just the sheaf of all maps from B to the $\mathcal{O}^\times$-gerbe on X whose characteristic class is $T_0(\alpha)$. Note that this gerbe is well defined in view of Corollary 2.15. We will write $g(\alpha) \in \mathrm{Ext}^1_{\mathbb{Z}_B}(\mathscr{X}, \mathcal{O}_B^\times[1]) = \mathrm{Ext}^2_{\mathbb{Z}_B}(\mathscr{X}, \mathcal{O}_B^\times)$ for the extension class of ${}_\alpha\mathscr{X}$. For more on the relevance of commutative group stacks see Section 1.

With this notation it is now clear that $\langle \alpha, \beta \rangle$ is just the Yoneda product $g(\beta) \circ e(\alpha)$.

LEMMA 2.16. *The bilinear pairing*
$$\langle \bullet, \bullet \rangle : H^1_{an}(B, \mathscr{X}) \bigotimes_{\mathbb{Z}} H^1_{an}(B, \mathscr{X}) \to H^3_{an}(B, \mathcal{O}_B^\times)$$
is skew-symmetric.

Proof. The Poincare sheaf $\mathscr{P} \to X \times_B X$ satisfies the biextension property and so can be interpreted functorially (see [**SGA7-I**, Exposé VII,Corollary 3.6.5]) as an object $\mathfrak{L}(\mathcal{P}) \in \mathrm{ob}\, D^b(\mathbb{Z}_B\text{-mod})$ in the derived category of abelian sheaves on B, which is an extension of $\mathscr{X} \overset{L}{\otimes} \mathscr{X}$ by $\mathcal{O}_B^\times$. In other words, $\mathfrak{L}(\mathcal{P})$ fits in a distinguished triangle
$$\mathcal{O}_B^\times \to \mathfrak{L}(\mathcal{P}) \to \mathscr{X} \overset{L}{\otimes} \mathscr{X} \to \mathcal{O}_B^\times[1]$$
of complexes of abelian sheaves. Let $\mathfrak{p} \in \mathrm{Ext}^1_{\mathbb{Z}_B}(\mathscr{X} \overset{L}{\otimes} \mathscr{X}, \mathcal{O}_B^\times) = \mathrm{Hom}(\mathscr{X} \overset{L}{\otimes} \mathscr{X}, \mathcal{O}_B^\times[1])$ be the corresponding extension class. From the definition of the homomorphisms
$$e(\alpha) \in \mathrm{Hom}_{D^b(\mathbb{Z}_B\text{-mod})}(\mathbb{Z}_B, \mathscr{X}[1]), \text{ and}$$
$$g(\alpha) \in \mathrm{Hom}_{D^b(\mathbb{Z}_B\text{-mod})}(\mathscr{X}, \mathcal{O}_B^\times[2])$$
one can easily check that $g(\alpha)$ can be identified with the composition
$$\mathscr{X} = \mathbb{Z}_B \otimes \mathscr{X} \xrightarrow{e(\alpha)\otimes \mathrm{id}_{\mathscr{X}}} \mathscr{X} \overset{L}{\otimes} \mathscr{X}[1] \xrightarrow{\mathfrak{p}} \mathcal{O}_B^\times[1].$$
Indeed, observe that both $g(\alpha)$ and $\mathfrak{p} \circ (e(\alpha) \otimes \mathrm{id}_{\mathscr{X}})$ can naturally be interpreted as amplitude one objects in the derived category of abelian

sheaves on B. Since any amplitude one object in $D^b(\mathbb{Z}_B\text{-mod})$ can be viewed as a stack over B it suffices to show the equivalence of the categories fibered in groupoids corresponding to $g(\alpha)$ and $\mathfrak{p} \circ (e(\alpha) \otimes \mathrm{id}_{\mathscr{X}})$ respectively. Let as before ${}_\alpha\mathscr{X} \to B$ denote the fibered category corresponding to $g(\alpha)$. Since by construction ${}_\alpha\mathscr{X}$ comes from the push-forward $R\pi_{\alpha *}\mathcal{O}^{\times}_{X_\alpha}$ we can identify explicitly the groupoid of sections of ${}_\alpha\mathscr{X}$ over an open set U in B as the groupoid of all line bundles L on $(X_\alpha \times_B X)_{|U}$ having the property that for any point $b \in U$ and any $x \in X_b$ we have that $L_{|(X_\alpha)_b \times \{x\}} \cong \mathcal{O}_{X_b}(x - \sigma(b))$. Finally, using the description of the complex $\mathfrak{L}(\mathscr{P})$ in terms of fibered categories given in [**SGA7-I**, Exposè VII] we see immediately that this groupoid is precisely the groupoid of sections over U of the fibered category corresponding to $\mathfrak{p} \circ (e(\alpha) \circ \mathrm{id}_{\mathscr{X}})$.

Now taking into account that $g(\alpha) = \mathfrak{p}\circ(e(\alpha)\otimes\mathrm{id}_{\mathscr{X}})$, we see that for any two elements $\alpha, \beta \in H^1_{an}(B, \mathscr{X})$ the product $\langle \alpha, \beta \rangle \in H^3_{an}(B, \mathcal{O}^{\times}_B)$ can be rewritten as the Yoneda product

$$\langle \alpha, \beta \rangle = \mathfrak{p} \circ (\alpha \Cup \beta),$$

where $\alpha \Cup \beta \in H^2_{an}(B, \mathscr{X} \overset{L}{\otimes} \mathscr{X})$ is the external cup product of α and β.

To understand the symmetry properties of $\langle \bullet, \bullet \rangle$ it only remains to notice that

$$\langle \beta, \alpha \rangle = \mathfrak{p} \circ (\beta \Cup \alpha) = \mathfrak{p} \circ \boldsymbol{sw}(\alpha \Cup \beta),$$

where $\boldsymbol{sw} : \mathscr{X} \overset{L}{\otimes} \mathscr{X} \to \mathscr{X} \overset{L}{\otimes} \mathscr{X}$ is the involution switching the two factors. However recall that $\mathscr{P}$ is a normalized Poincare bundle and so can be explicitly described as the rank one divisorial sheaf

$$\mathscr{P} = \mathcal{O}_{X \times_B X}(\Delta - \sigma \times_B X - X \times_B \sigma - \varpi^* N_{\sigma/X})$$

where $\varpi : X \times_B X \to B$ is the natural projection and $N_{\sigma/X}$ is the normal bundle to the section $\sigma \subset X$. In particular $\boldsymbol{sw}^*(\mathscr{P}) = \mathscr{P}$ and so $\mathscr{P} \to X \times_B X$ is a symmetric biextension. This shows that $\mathfrak{p} \circ \boldsymbol{sw} = \mathfrak{p}$. Combined with the fact that $\boldsymbol{sw}(\beta \Cup \alpha) = (-1)^{|\alpha|\cdot|\beta|}\alpha \Cup \beta = -\alpha \Cup \beta$ we conclude that $\langle \beta, \alpha \rangle = -\langle \alpha, \beta \rangle$. The lemma is proven. □

An immediate corollary of the skew-symmetry of $\langle \bullet, \bullet \rangle$ is that $T_\alpha(\beta) \in Br'_{an}(X_\alpha)$ is well defined iff $T_\beta(\alpha) \in Br'_{an}(X_\beta)$ is well defined.

In the case of surfaces we get the following:

COROLLARY 2.17. *Suppose that X is a smooth surface and that π is non-isotrivial with all fibers integral. Then $\text{Ш}(X)$ is infinitely* *divisible and so any* $\alpha \in Ш_{an}(X)$ *is* m*-compatible with all elements in* $Ш_{an}(X)[m]$.

Proof. To show that $Ш(X)$ is infinitely divisible note that since B is a curve we can apply Corollary 2.15 to conclude that the map T_0 fits in a short exact sequence

$$0 \to Ш_{an}(X) \to Br'_{an}(X) \to H^1(B, \mathbb{Z})$$

where the last map is the composition of the identification $Br'_{an}(X) \cong H^1_{an}(B, \mathscr{P}ic(X/B))$ coming from the Leray spectral sequence and the map $H^1_{an}(B, \mathscr{P}ic(X/B)) \to H^1(B, \mathbb{Z})$ corresponding to the degree morphism $\deg : \mathscr{P}ic(X/B) \to \mathbb{Z}_B$.

Since by assumption π has only integral fibers, we have natural identifications

$$\mathscr{P}ic(X/B) = R^1\pi_*\mathcal{O}_X^\times, \quad \text{and} \quad \mathbb{Z}_B = R^2\pi_*\mathbb{Z}_X$$

under which the degree map $\deg : \mathscr{P}ic(X/B) \to \mathbb{Z}_B$ becomes the coboundary homomorphism $\delta : R^1\pi_*\mathcal{O}_X^\times \to R^2\pi_*\mathbb{Z}_X$ in the long exact sequence of higher direct images associated to the exponential sequence

$$0 \to \mathbb{Z}_X \to \mathcal{O}_X \overset{\exp}{\to} \mathcal{O}_X^\times \to 1$$

and the map $\pi : X \to B$.

In particular, this implies that the map $Br'_{an}(X) \to H^1(B, \mathbb{Z})$ fits in the commutative diagram:

$$\begin{array}{ccc}
Br'_{an}(X) & \longrightarrow & H^1(B, \mathbb{Z}) \\
\downarrow \cong & & \cong \downarrow \\
H^1_{an}(B, R^1\pi_*\mathcal{O}_X^\times) & \xrightarrow{\delta} & H^1(B, R^2\pi_*\mathbb{Z}_X) \\
\uparrow \theta & & \eta \uparrow \\
H^2_{an}(X, \mathcal{O}_X^\times) & \xrightarrow{\delta} & H^3(X, \mathbb{Z}).
\end{array}$$

Here the maps θ and η between the third and second rows come from the Leray spectral sequences for the map $\pi : X \to B$ and the sheaves $\mathcal{O}_X^\times$ and $\mathbb{Z}_X$, which give:

$$H^2_{an}(B, \mathcal{O}_B^\times) \longrightarrow H^2_{an}(X, \mathcal{O}_X^\times) \xrightarrow{\theta} H^1_{an}(B, R^1\pi_*\mathcal{O}_X^\times) \longrightarrow 0$$

and

$$H^2(B, R^1\pi_*\mathbb{Z}_X) \longrightarrow H^3(X, \mathbb{Z}) \xrightarrow{\eta} H^1(B, R^2\pi_*\mathbb{Z}_X) \longrightarrow 0.$$

Now $H^2_{an}(B, \mathcal{O}_B^\times) = 0$ since B is one dimensional, and $H^2(B, R^1\pi_*\mathbb{Z}_X) = 0$ by the irreducibility of monodromy. This implies that

$$\begin{aligned} \text{Ш}_{an}(X) &= \ker[Br'_{an}(X) \to H^1(B, \mathbb{Z})] \\ &= \text{im}[H^2_{an}(X, \mathcal{O}_X) \to H^2_{an}(X, \mathcal{O}_X^\times)] \end{aligned}$$

and so $\text{Ш}(X)$ is divisible. The corollary is proven. □

DEFINITION 2.18. For any complementary pair $\alpha, \beta \in \text{Ш}_{an}(X)$ we denote the $\mathcal{O}^\times$-gerbe on X_β classified by $T_\beta(\alpha)$ by ${}_\alpha X_\beta$.

CONJECTURE 2.19. For any complementary pair $\alpha, \beta \in \text{Ш}_{an}(X)$, there exists an equivalence

$$D^b_1({}_\alpha X_\beta) \cong D^b_{-1}({}_\beta X_\alpha)$$

of the bounded derived categories of sheaves of pure weights ± 1 on ${}_\alpha X_\beta$ and ${}_\beta X_\alpha$ respectively.

In section 4 we will prove this conjecture in any dimension under the additional assumptions that π is smooth and that α and β are m-compatible. In chapter 4 we will prove it unconditionally when X is a surface.

CHAPTER 3

Smooth genus one fibrations

In this section we will consider smooth genus one fibrations over smooth bases of arbitrary dimension and $\mathcal{O}^{\times}$-gerbes over them.

1. $\mathcal{O}^{\times}$-gerbes

In this section we work with a fixed smooth elliptic fibration

$$X \underset{\sigma}{\overset{\pi}{\rightleftarrows}} B,$$

and two genus one fibrations X_α, X_β corresponding to two m-compatible elements $\alpha, \beta \in \text{Ш}_{an}(X)$. Recall that m-compatibility means that one of the elements, say β, is actually algebraic, i.e. β is a torsion element of some order m in $\text{Ш}_{an}(X)$, while α is an m-divisible element. Choose an element $\varphi \in \text{Ш}_{an}(X)$ such that $m\varphi = \alpha$. We will use this data to construct presentations for gerbes ${}_\beta\mathscr{E}_\alpha$ over X_α and ${}_\alpha\mathscr{L}_\beta$ over X_β. Different choices of the root φ give rise to different but Morita equivalent presentations of the same gerbes.

1.1. The lifting presentation. Recall from chapter 2 that a gerbe presentation over a variety X is a diagram

$$\begin{array}{ccccc} R & & \overset{s}{\underset{t}{\searrow\!\!\searrow}} & & \\ \downarrow & & & & \\ U \times_X U & \overset{p_1}{\underset{p_2}{\rightrightarrows}} & U & \xrightarrow{u} & X \end{array}$$

where $R \to U \times_X U$ is a $\mathbb{C}^{\times}$-bundle satisfying the biextension condition.

We define a gerbe ${}_\alpha\mathscr{L}_\beta$ on X_β via the *lifting presentation*:

(3.1)

$$\begin{array}{ccccc} {}_\alpha LR_\beta & & \overset{s}{\underset{t}{\searrow\!\!\searrow}} & & \\ \downarrow & & & & \\ {}_\alpha LU_\beta \times_{X_\beta} {}_\alpha LU_\beta & \overset{p_1}{\underset{p_2}{\rightrightarrows}} & {}_\alpha LU_\beta & \xrightarrow{p_2} & X_\beta \end{array}$$

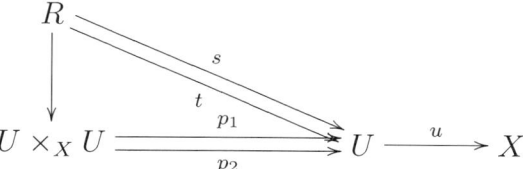

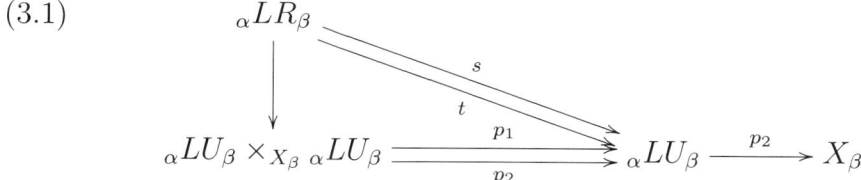

where

$$
\begin{aligned}
{}_\alpha LU_\beta &:= X_\varphi \times_B X_\beta,\\
p_2 : {}_\alpha LU_\beta = X_\varphi \times_B X_\beta \to X_\beta &\quad \text{is the second projection, and}\\
{}_\alpha LR_\beta &:= \operatorname{tot}(\mathscr{P}^\times_{1-2,\, m\cdot 3}) \to ({}_\alpha LU_\beta) \times_{X_\beta} ({}_\alpha LU_\beta).
\end{aligned}
$$

Here $\mathscr{P}_{1-2,\, m\cdot 3} \to ({}_\alpha LU_\beta) \times_{X_\beta} ({}_\alpha LU_\beta)$ is the pullback via the map

$$
\begin{aligned}
p_{1-2,\, m\cdot 3} : \qquad X_\varphi \times_B X_\varphi \times_B X_\beta &\longrightarrow X \times_B X\\
(a, b, x) &\longmapsto (a - b, m \cdot x),
\end{aligned}
$$

of the Poincare bundle

$$\mathscr{P} := \mathcal{O}_{X \times_B X}(\Delta - \sigma \times_B X - X \times_B \sigma - \varpi^* c_1(B))$$

on $X \times_B X$. As usual we denote the natural projection $X \times_B X \to B$ by ϖ. The required biextension property for $\mathscr{P}_{1-2,\, m\cdot 3}$ follows immediately from the see-saw principle .

For future reference we note that under the obvious identification

$$({}_\alpha LU_\beta) \times_{X_\beta} ({}_\alpha LU_\beta) = X_\varphi \times_B X_\varphi \times_B X_\beta$$

the lifting presentation (3.1) can be rewritten as

(3.2)
$$
\begin{array}{ccccc}
\operatorname{tot}(\mathscr{P}^\times_{1-2,\, m\cdot 3}) & \overset{s}{\underset{t}{\rightrightarrows}} & & & \\
\downarrow & & & & \\
X_\varphi \times_B X_\varphi \times_B X_\beta & \overset{p_{13}}{\underset{p_{23}}{\rightrightarrows}} & X_\varphi \times_B X_\beta & \xrightarrow{p_2} & X_\beta.
\end{array}
$$

1.2. The extension presentation. Similarly, we define a gerbe ${}_\beta\mathscr{E}_\alpha$ on X_α via the *extension presentation*:

(3.3)
$$
\begin{array}{ccccc}
{}_\beta ER_\alpha & \overset{s}{\underset{t}{\rightrightarrows}} & & & \\
\downarrow & & & & \\
{}_\beta EU_\alpha \times_{X_\alpha} {}_\beta EU_\alpha & \overset{p_1}{\underset{p_2}{\rightrightarrows}} & {}_\beta EU_\alpha & \xrightarrow{q_\varphi} & X_\alpha
\end{array}
$$

where

$$
\begin{aligned}
{}_\beta EU_\alpha &:= X_\varphi,\\
q_\varphi : X_\varphi \to X_\alpha &\quad \text{is the multiplication by } m \text{ map, and}\\
{}_\beta ER_\alpha &:= \operatorname{tot}(\Phi^\times_\beta) \to X_\varphi \times_{X_\alpha} X_\varphi.
\end{aligned}
$$

Here Φ_β could be taken as the line bundle $d^*(M)$ on $X_\varphi \times_{X_\alpha} X_\varphi$, where

$$d : X_\varphi \times_{X_\alpha} X_\varphi \to X[m]$$

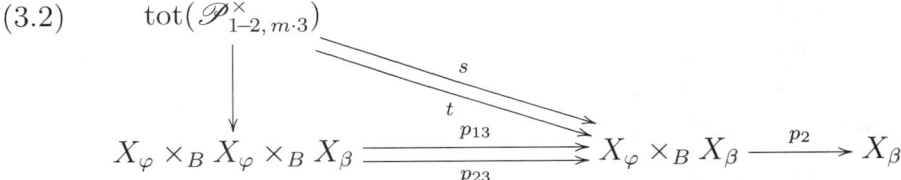

is the difference map and M is any line bundle on $X[m]$ whose punctured total space gives a group extension:

$$1 \longrightarrow \mathcal{O}_B^\times \longrightarrow \mathrm{tot}(M^\times) \longrightarrow X[m] \longrightarrow 0.$$

We will explain below how to construct a relative line bundle $\Sigma_\beta \in \Gamma(B, \mathscr{P}ic^m(X_\beta/B))$ and a global line bundle $M_\beta \to X[m]$, determined by the condition that its punctured total space is the theta group G_β:

$$1 \longrightarrow \mathcal{O}_B^\times \longrightarrow G_\beta \longrightarrow X[m] \longrightarrow 0$$

of Σ_β. The simplest choice would be to take $M := M_\beta$. However, we will see later that in order to achieve duality with the lifting gerbe, the correct choice is to take $M := M_\beta \otimes M_0^{-1}$, where M_0 is determined by the condition that its punctured total space is the theta group G_0:

$$1 \longrightarrow \mathcal{O}_B^\times \longrightarrow G_0 \longrightarrow X[m] \longrightarrow 0$$

corresponding to the similarly defined relative line bundle $\Sigma_0 \in \Gamma(B, \mathscr{P}ic^m(X/B))$.

To define Σ_β, consider first two genus one curves E' and E'' with the same Jacobian E. Let $q : E' \to E''$ be a map which induces the multiplication by m map $E \to E$. For any two points $a, b \in E'$ such that $q(a) = q(b)$, we have that $\mathcal{O}_{E'}(m \cdot a) \cong \mathcal{O}_{E'}(m \cdot b)$. This determines a map $E'' \to \mathrm{Pic}^m(E')$. Applying this to our map $q_\beta : X_\beta \to X$ we get a well defined morphism $X \to \mathrm{Pic}^m(X_\beta/B)$. The image of $\sigma \subset X$ is our relative line bundle Σ_β. Similarly we get Σ_0 from the multiplication by m map $q_0^m : X \to X$. Note that by construction the relative line bundle $\Sigma_0 \in \mathrm{Pic}^m(X/B)$ is induced by a global line bundle $\mathcal{O}_X(m\sigma)$. Similarly, if $Br'_{an}(B) = 0$, then the relative line bundle $\Sigma_\beta \in \mathrm{Pic}^m(X/B)$ is induced by some global line bundle $\mathcal{O}_{X_\beta}(\sigma_\beta)$, where $\sigma_\beta \subset X_\beta$ is an m-section of π_β.

Given any relative line bundle $\Sigma \in \Gamma(B, \mathrm{Pic}^m(Y/B))$, where $\pi_Y : Y \to B$ is in Ш(X), we get a global line bundle M on $X[m]$ equipped with an isomorphism

$$\mu^* M \cong \mathrm{pr}_1^* M \otimes \mathrm{pr}_2^* M$$

satisfying the usual biextension property. Here $\mu : X[m] \times_B X[m] \to X[m]$ is multiplication, and pr_1, pr_2 are the projections.

Locally on B, choose an actual line bundle $\Sigma' \in \mathrm{Pic}(Y)$ lifting Σ. We define $M' := p_{1*}(a^*\Sigma' \otimes p_2^*\Sigma'^{-1})$. Here $a : X[m] \times_B Y \to Y$ is the action and p_1, p_2 are the projections. Note that $a^*\Sigma' \otimes p_2^*\Sigma'^{-1}$ is trivial on each fiber of p_1, so M' is a line bundle. Let M'' be the line bundle determined by another lift Σ'' of Σ. Any such Σ'' is necessarily of the form $\Sigma' \otimes \pi_Y^* L$ for some line bundle L on B. The equality $\pi_Y \circ a = \pi_Y \circ p_2$ therefore induces a canonical isomorphism

$$a^*\Sigma' \otimes p_2^*\Sigma'^{-1} \to a^*\Sigma'' \otimes p_2^*\Sigma''^{-1},$$

so the locally defined line bundles M', M'' glue to a global line bundle M on $X[m]$.

The biextension property of M is equivalent to saying that its punctured total space $G := \mathrm{tot}(M^\times)$ has the structure of a group scheme over B called the *theta group* of Σ. It is a central extension of $X[m]$ by $\mathbb{G}_m$. Explicitly a local section of G is a pair (x, λ), where x is a local section of $X[m] \to B$ and $\lambda : t_x^*\Sigma \to \Sigma$ is an isomorphism.

Applying these constructions to our relative line bundles Σ_β and Σ_0 produces the desired line bundles M_β and M_0 and theta groups G_β and G_0.

For future reference we note that under the obvious isomorphism

$$d \times p_2 : X_\varphi \times_{X_\alpha} X_\varphi \widetilde{\to} X[m] \times_B X_\varphi$$

the extension presentation (3.3) can be rewritten in the equivalent form:

(3.4)
$$\begin{array}{ccccc} \mathrm{tot}(\Phi_\beta^\times) & \overset{s}{\underset{t}{\rightrightarrows}} & X_\varphi & & \\ \downarrow & & & & \\ X[m] \times_B X_\varphi & \overset{a_\varphi}{\underset{p_2}{\rightrightarrows}} & X_\varphi & \xrightarrow{q_\varphi} & X_\alpha, \end{array}$$

where $a_\varphi : X[m] \times_B X_\varphi \to X_\varphi$ denotes the action and p_2 denotes the second projection.

1.3. Coboundary realizations. Although we constructed ${}_\alpha\mathscr{L}_\beta$ and ${}_\beta\mathscr{E}_\alpha$ directly, it is worth noting that the lifting and extension presentations are both special cases of the coboundary construction described in Sections 1.1 and 1.2. Recall that the input for the coboundary construction for a gerbe presentation on a variety Y consists of a short exact sequence of group schemes over Y

$$1 \to \mathbb{G}_m \to G \to K \to 1$$

together with a K torsor U.

The lifting presentation is obtained from the short exact sequence

$$1 \to \mathbb{G}_m \to \mathrm{tot}(p_{1,m\cdot 2}^*\mathscr{P}^\times) \to \pi_\beta^*(X) \to 1$$

and the $\pi_\beta^*(X) = X \times_B X_\beta$-torsor $X_\varphi \times_B X_\beta$. Note that the group structure on $\mathrm{tot}(p_{1,m\cdot 2}^*\mathscr{P}^\times)$ in the above sequence comes from the biextension property of the Poincare bundle (for the group law on $\pi_\beta^*(X)$).

The extension presentation is obtained from the short exact sequence

$$1 \to \mathbb{G}_m \to \operatorname{tot}(\Phi_\beta^\times) \to \pi_\alpha^*(X[m]) \to 1$$

and the $\pi_\alpha^*(X[m]) = X[m] \times_B X_\alpha$-torsor X_φ.

In other words, we can write ${}_\alpha\mathscr{L}_\beta$ and ${}_\beta\mathscr{E}_\alpha$ as quotient gerbes:

$$\begin{aligned} {}_\alpha\mathscr{L}_\beta &= [X_\varphi \times_B X_\beta / \operatorname{tot}(p_{1,m\cdot 2}^* \mathscr{P}^\times)] \\ {}_\beta\mathscr{E}_\alpha &= [X_\varphi / \operatorname{tot}(\Phi_\beta^\times)]. \end{aligned}$$

2. The class of the lifting gerbe

In this section we continue to assume that $\pi : X \to B$ is smooth, $\beta \in Ш_{an}(X)$ is of finite order m and $\alpha \in Ш_{an}(X)$ is m-divisible.

THEOREM 3.1. *The class $[{}_\alpha\mathscr{L}_\beta]$ of the lifting gerbe equals $T_\beta(\alpha)$. In other words ${}_\alpha\mathscr{L}_\beta$ is a model for ${}_\alpha X_\beta$.*

Proof. The proof is in two steps. In step (1) we show that a cocycle representing the class $T_\beta(\alpha)$ which defines ${}_\alpha X_\beta$ becomes a coboundary $\delta(c)$ when pulled back to $L := {}_\alpha LU_\beta$. In step (2) we check that the line bundle defined by c on $L \times_{X_\beta} L$ coincides with the Poincare bundle ${}_\alpha LR_\beta$.

We need to show the isomorphism of two gerbes on the smooth space X_β.

We will be working with the Cartesian product

$$\begin{array}{ccc} L & \xrightarrow{\lambda_\varphi} & X_\varphi \\ {\scriptstyle\lambda_\beta}\downarrow & \searrow^{\lambda_B} & \downarrow{\scriptstyle\pi_\varphi} \\ X_\beta & \xrightarrow[\pi_\beta]{} & B \end{array}$$

Recall from section 1 that the class of ${}_\alpha X_\beta$ is $T_\beta(\alpha) \in H^2(X_\beta, \mathscr{O}^\times)$. By the Leray spectral sequence for $\pi_\beta : X_\beta \to B$, this group equals $H^1(B, \mathscr{P}ic(X_\beta/B))$. Explicitly, $T_\beta(\alpha)$ is the class of the $\mathscr{P}ic(X_\beta/B)$-torsor induced from the $\mathscr{P}ic^0(X_\beta/B) = X$-torsor X_β.

Similarly, the Leray spectral sequence for $\lambda_B : L \to B$ gives an injection

$$H^1(B, \mathscr{P}ic(L/B)) \hookrightarrow H^2(L, \mathscr{O}^\times). \tag{3.5}$$

The λ_β-pullback of $T_\beta(\alpha)$ is in the image of (3.5) and is the $\mathscr{P}ic(L/B)$-torsor induced from X_β via $\lambda_\beta^* : \mathscr{P}ic^0(X_\beta/B) \to \mathscr{P}ic(L/B)$.

For step (1), consider the short exact sequence of sheaves of groups on B:

$$0 \to \mathscr{P}ic(X_\beta/B) \xrightarrow{\lambda_\beta^*} \mathscr{P}ic(L/B) \xrightarrow{\text{ev}} \mathscr{H}om_B(X_\beta, \operatorname{Pic}(X_\varphi/B)) \to 0.$$

Here ev sends a line bundle on L to the family of its restrictions on $\text{pt} \times_B X_\varphi$, and it is surjective because $\pi_\beta : X_\beta \to B$ is smooth.

We are claiming that $\lambda_\beta^*(T_\beta(\alpha)) = 0$, so we need to show that $T_\beta(\alpha)$ is in the image of the coboundary

$$\partial : H^0(B, \mathscr{H}om_B(X_\beta, \operatorname{Pic}(X_\varphi/B))) \to H^1(B, \mathscr{P}ic(X_\beta/B)).$$

In $H^0(B, \mathscr{H}om_B(X_\beta, \operatorname{Pic}(X_\varphi/B)))$ we have a natural element

$$q : X_\beta \to X_{m\beta} = X = \operatorname{Pic}^0(X_\varphi/B) \subset \operatorname{Pic}(X_\varphi/B),$$

depending on the choice of a trivialization Σ_β of $X_{m\beta}$.

We will see that $\partial(q) = T_\beta(\alpha)$.

Let $\mathfrak{U} = \{U_i\}_{i\in I}$ be an analytic open cover of B for which we have trivializations $s_i : U_i \to X_\varphi$ of the X-torsor X_φ. In order to calculate $\partial(q)$ we first lift q to an element $c \in C^0(\mathfrak{U}, \mathscr{P}ic(L/B))$. This lift is given in terms of the map

$$t_{-s_i} \times q : L_{|U_i} = X_\varphi \times_{U_i} X_\beta \to X \times_{U_i} X$$

by $c_i := (t_{-s_i} \times q)^* \mathscr{P}$, where $\mathscr{P}$ is the standard Poincare bundle on $X \times_B X$.

The Čech differential $\delta(c) \in Z^1(\mathfrak{U}, \mathscr{P}ic(L/B))$ is given by $\{c_i \otimes c_j^{-1}\}_{i,j\in I}$. It comes from $Z^1(\mathfrak{U}, \mathscr{P}ic(X_\beta/B))$, and is represented there by $\{\mathcal{O}_{\pi_\beta^{-1}(U_{ij})}(m(s_j - s_i))\}_{i,j\in I}$. On the other hand, ms_i can be interpreted as a section of $X_\alpha = X_{m\varphi}$ over U_i, so this cocycle represents our $T_\beta(\alpha)$.

For step (2), consider the cochain

$$\{p_1^* c_i \otimes p_2^* c_i^{-1}\}_{i\in I} \in C^0(\mathfrak{U}, \mathscr{P}ic(L \times_{X_\beta} L/B)).$$

¿From the discussion in section 1 we know that this cochain is in fact a global section $\mathbb{L}$ of $\mathscr{P}ic(L \times_{X_\beta} L/B)$. We need to show that $\mathbb{L} = \mathscr{P}_{1-2,\, m\cdot 3}$. As usual we identify $L \times_{X_\beta} L$ with $X_\varphi \times_B X_\varphi \times_B X_\beta$, so p_1 and p_2 become p_{13} and p_{23}.

It suffices to show the equality $\mathbb{L}_{|\lambda_B^{-1}(U_i)} = p_1^* c_i \otimes p_2^* c_i^{-1}$ for each open set U_i. This follows by the theorem of the cube from the identifications:

$$\begin{aligned} p_1^* c_i &= p_{13}^* \circ (t_{-s_i} \times q)^* \mathscr{P} \\ p_2^* c_i &= p_{23}^* \circ (t_{-s_i} \times q)^* \mathscr{P} \\ \mathbb{L} &= p_{1-2,\, m\cdot 3}^* \mathscr{P}. \end{aligned}$$

This finishes the proof of the theorem. □

3. The class of the extension gerbe

In this section we again assume that $\pi : X \to B$ is a smooth elliptic fibration, that $Br'_{an}(B) = 0$ and that $\alpha, \beta \in Ш_{an}(X)$ are m-compatible with β of order m.

THEOREM 3.2. *The class* $[{}_{\beta}\mathscr{E}_{\alpha}]$ *of the extension gerbe equals* $T_{\alpha}(\beta)$. *In other words* ${}_{\beta}\mathscr{E}_{\alpha}$ *is a model for* ${}_{\beta}X_{\alpha}$.

Proof. Recall that the assumption $Br'_{an}(B) = 0$ together with the Leray spectral sequence for $\pi_{\alpha} : X_{\alpha} \to B$ give us an injection

$$H^2_{an}(X_{\alpha}, \mathcal{O}^{\times}) \hookrightarrow H^1_{an}(B, R^1\pi_{\alpha *}\mathcal{O}^{\times}) = H^1(B, \mathscr{P}ic(X_{\alpha}/B)).$$

In terms of this inclusion, $T_{\alpha}(\beta)$ can be identified with the isomorphism class of the $\mathscr{P}ic(X_{\alpha}/B)$-torsor associated to the $\mathscr{P}ic^0(X_{\alpha}/B) = \mathscr{X}$ torsor X_{β}. In order to show that $[{}_{\beta}\mathscr{E}_{\alpha}] = T_{\alpha}(\beta)$ we must first check that $T_{\alpha}(\beta)$ pulls back to the trivial element in $H^2_{an}({}_{\beta}EU_{\alpha}, \mathcal{O}^{\times})$.

Recall from Section 1.2 that the atlas ${}_{\beta}EU_{\alpha}$ for the extension presentation is defined by fixing an element $\varphi \in Ш_{an}(X)$ such that $m \cdot \varphi = \alpha$ and then taking ${}_{\beta}EU_{\alpha} := X_{\varphi}$. With this definition the structure morphism ${}_{\beta}EU_{\alpha} \to X_{\alpha}$ is identified with the map $q := q^m_{\varphi} : X_{\varphi} \to X_{\alpha}$ of multiplication by m along the fibers.

The pullback via q of relative line bundles defined on the fibers of $\pi_{\alpha} : X_{\alpha} \to B$ gives rise to a morphism of sheaves of groups

$$Q : \mathscr{P}ic(X_{\alpha}/B) \longrightarrow \mathscr{P}ic(X_{\varphi}/B).$$

Since q corresponds to multiplication by m, it follows that Q fits in a commutative diagram

$$\begin{array}{ccc}
\mathscr{P}ic(X_{\alpha}/B) & \xrightarrow{\;Q\;} & \mathscr{P}ic(X_{\varphi}/B) \\
\Big\uparrow & & \Big\uparrow \\
\mathscr{P}ic^0(X_{\alpha}/B) & & \mathscr{P}ic^0(X_{\varphi}/B) \\
\Big\| & & \Big\| \\
\mathscr{X} & \xrightarrow{\;\mathrm{mult}_m\;} & \mathscr{X}
\end{array}$$

Also, since $\pi_{\alpha} \circ q = \pi_{\varphi}$, it follows that the pullback map

$$q^* : H^2_{an}(X_{\alpha}, \mathcal{O}^{\times}) \longrightarrow H^2_{an}(X_{\varphi}, \mathcal{O}^{\times})$$

is compatible with the Leray spectral sequences for π_α and π_φ, and so fits in a commutative diagram

$$\begin{array}{ccc}
H^2_{an}(X_\alpha, \mathcal{O}^\times) & \xrightarrow{q^*} & H^2_{an}(X_\varphi, \mathcal{O}^\times) \\
\big\downarrow & & \big\downarrow \\
H^1_{an}(B, \mathscr{P}ic(X_\alpha/B)) & \xrightarrow{h^1(Q)} & H^1_{an}(B, \mathscr{P}ic(X_\varphi/B)) \\
\big\uparrow & & \big\uparrow \\
H^1(B, \mathscr{X}) & \xrightarrow{h^1(\mathrm{mult}_m)} & H^1(B, \mathscr{X})
\end{array}$$

Thus we can identify $q^*(T_\alpha(\beta))$ with the class of the $\mathscr{P}ic(X_\varphi/B)$-torsor which is induced from the $\mathscr{P}ic^0(X_\varphi/B) = \mathscr{X}$-torsor $h^1(\mathrm{mult}_m)(X_\beta) = X_{m\cdot\beta}$. However by assumption $m \cdot \beta = 0$ and so $X_{m\cdot\beta}$ is trivial as a $\mathscr{X}$-torsor. Therefore $q^*(T_\alpha(\beta)) = 0$ as promised.

To complete the proof of the theorem we need to realize the cocycle $q^*(T_\alpha(\beta))$ as a coboundary:

$$q^*(T_\alpha(\beta)) = \partial(\psi)$$

for some $\psi \in C^1_{an}(X_\varphi, \mathcal{O}^\times)$, and then check that the line bundle defined by ψ on $X_\varphi \times_{X_\alpha} X_\varphi$ is isomorphic to Φ_β.

In terms of the inclusion $H^2_{an}(X_\varphi, \mathcal{O}^\times) \subset H^1_{an}(B, \mathscr{P}ic(X_\varphi/B))$ this amounts to writing the class $q^*(T_\alpha(\beta)) \in H^1_{an}(B, \mathscr{P}ic(X_\varphi/B))$ as the coboundary of some Čech cochain $\psi \in C^0_{an}(b, \mathscr{P}ic(X_\varphi/B))$ and then showing that the global section of $\mathscr{P}ic(X_\varphi \times_{X_\alpha} X_\varphi/B)$ determined by ψ coincides with the global section given by Φ_β. To carry this out we will need to first choose a cocycle representating of $T_\beta(\alpha) \in H^1_{an}(B, \mathscr{P}ic(X_\alpha/B))$ or equivalently a cocycle representating for the $\mathscr{X}$-torsor X_β.

Let $\mathfrak{U} = \{U_i\}$ be an analytic open covering of B which trivializes X_β as an X torsor. Choose trivializing sections $s_i \in \Gamma(U_i, X_\beta)$ over each U_i. Then $T_\alpha(\beta) \in H^1_{an}(B, \mathscr{P}ic(X_\alpha/B))$ is represented by the Čech cocycle

$$\begin{array}{ll}
\{\mathcal{O}_{X_\beta}(s_j - s_i)\} \in Z^1(\mathfrak{U}, \mathscr{X}) & \\
\qquad\qquad\qquad\quad \| & \\
\qquad\qquad\qquad Z^1(\mathfrak{U}, \mathscr{P}ic^0(X_\alpha/B)) \to Z^1(\mathfrak{U}, \mathscr{P}ic(X_\alpha/B)). &
\end{array}$$

Here $\mathcal{O}_{X_\beta}(s_j - s_i)$ is viewed as a line bundle of degree zero along the fibers of $\pi_\alpha : X_{\alpha|U_{ij}} \to U_{ij}$ via the canonical identification

$$\mathscr{P}ic^0(X_\beta/B) = \mathscr{X} = \mathscr{P}ic^0(X_\alpha/B).$$

In particular $q^*(T_\alpha(\beta)) \in H^1_{an}(B, \mathscr{P}ic(X_\varphi/B))$ is represented by the cocycle

$$\begin{array}{c} \{\mathcal{O}_{X_\beta}(s_j - s_i)^{\otimes m}\} \in Z^1(\mathfrak{U}, \mathscr{X}) \\ \| \\ Z^1(\mathfrak{U}, \mathscr{P}ic^0(X_\varphi/B)) \to Z^1(\mathfrak{U}, \mathscr{P}ic(X_\varphi/B)). \end{array}$$

In order to write this cocycle as a coboundary we will have to trivialize the $\mathscr{X}$-torsor $X_{m\cdot\beta}$.

Recall that in the construction of the line bundle Φ_β we used a particular trivialization of $X_{m\cdot\beta}$, namely the relative line bundle

$$\Sigma_\beta \in \Gamma_{an}(B, \mathscr{P}ic^m(X_\beta/B)).$$

Using Σ_β we can construct a cochain

$$\psi = \{\psi_i\} \in C^0(\mathfrak{U}, \mathscr{P}ic^0(X_\varphi/B))$$

with $\psi_i := \mathcal{O}_{X_\beta}(-m \cdot s_i) \otimes \Sigma_\beta$. By construction, ψ determines a global section

$$p_1^*\psi \otimes p_2^*\psi^{-1} : B \to \mathrm{Pic}^0(X_\varphi \times_{X_\alpha} X_\varphi/B),$$

namely, the section locally given by

$$p_1^*\psi_i \otimes p_2^*\psi_i^{-1} \in \Gamma_{an}(U_i, \mathscr{P}ic^0(X_\varphi \times_{X_\alpha} X_\varphi/B)).$$

On the other hand the section corresponding to $\Phi_\beta \to X_\varphi \times_{X_\alpha} X_\varphi$ can be described as follows.

Recall from section 1.2 that

$$\Phi_\beta = d^*(M_\beta \otimes M_0^{-1})$$

where $d : X_\varphi \times_{X_\alpha} X_\varphi \to X[m]$ is the natural difference map and M_β, M_0 are line bundles on $X[m]$ satisfying

$$\begin{aligned} d_\beta^* M_\beta &\cong p_1^*\Sigma_\beta \otimes p_2^*\Sigma_\beta^{-1} \\ d_0^* M_0 &\cong p_1^*\Sigma_0 \otimes p_2^*\Sigma_0^{-1}. \end{aligned}$$

Here again

$$\begin{array}{c} X_\beta \underset{q_\beta^m, X, q_\beta^m}{\times} X_\beta \xrightarrow{d_\beta} X[m] \\ X \underset{q_0^m, X, q_0^m}{\times} X \xrightarrow{d_0} X[m] \end{array}$$

stand for the difference maps.

By construction $p_1^*\psi_i \otimes p_2^*\psi_i^{-1}$ lives naturally in $\Gamma(U_i, \mathscr{P}ic^0(X_\beta \times_B X_\beta/B))$ (which we have identified with $\Gamma(U_i, \mathscr{P}ic^0(X_\varphi \times_B X_\varphi/B))$). In view of this, it will be convenient if we rewrite all objects as line bundles on $X_\beta \times_X X_\beta$. To that end, choose a local section $s : U_i \to X_\beta$ and let t_{-s} be the induced isomorphism

$$\begin{array}{ccc} X_{\beta|U_i} & \xrightarrow{t_{-s}} & X_{|U_i} \\ & \searrow \quad \swarrow & \\ & U_i & \end{array}$$

of translation by s along the fibers. With this notation we have a commutative diagram

$$\begin{array}{ccccc} X[m]_{|U_i} & \xleftarrow{d_\beta} & (X_\beta \times_X X_\beta)_{|U_i} & \overset{p_1}{\underset{p_2}{\rightrightarrows}} & X_{\beta|U_i} \\ \| & & \downarrow{\scriptstyle t_{-s}\times t_{-s}} & & \downarrow{\scriptstyle t_{-s}} \\ X[m]_{|U_i} & \xleftarrow{d_0} & (X \times_X X)_{|U_i} & \overset{p_1}{\underset{p_2}{\rightrightarrows}} & X_{|U_i} \end{array}$$

and thus

$$d_\beta^* M_0 = (t_s \times t_{-s})^* \circ d_0^* M_0 = p_1^* \mathcal{O}_{X_\beta}(ms) \otimes p_2^* \mathcal{O}_{X_\beta}(-ms).$$

Therefore, in order to compare $p_1^*\psi_i \otimes p_2^*\psi_i^{-1}$ and $d_\beta^*(M_\beta \otimes M_0^{-1})$, we need to show that on $(X_\beta \times_X X_\beta)_{|U_i}$ we have

$$p_1^*(\Sigma_\beta(-ms_i)) \otimes p_2^*(\Sigma_\beta(ms_i)) \cong p_1^*(\Sigma_\beta(-ms)) \otimes p_2^*(\Sigma_\beta(ms))$$

for every section $s : U_i \to X_\beta$.

Equivalently, it suffices to show that

$$p_1^* \mathcal{O}_{X_\beta}(m(s - s_i)) \otimes p_2^* \mathcal{O}_{X_\beta}(m(s_i - s)) \cong \mathcal{O}. \tag{3.6}$$

On the other hand we have a commutative diagram

$$\begin{array}{ccccc} & & X_\beta \times_X X_\beta & & \\ & \swarrow^{p_1} & & {}^{p_2}\searrow & \\ X_\beta & & \downarrow{\scriptstyle \mu} & & X_\beta \\ & {}_{q_\beta^m}\searrow & & \swarrow_{q_\beta^m} & \\ & & X & & \end{array}$$

and since $\mathcal{O}_{X_\beta}(m(s - s_i))$ is of degree zero along the fibers of π_β we have

$$\mathcal{O}_{X_\beta}(m(s - s_i)) = (q_\beta^m)^* \mathcal{O}_X(s - s_i). \tag{3.7}$$

Here $\mathcal{O}_X(s - s_i) \in \Gamma(U_i, \mathscr{P}ic^0(X/B))$ denotes the relative line bundle on X corresponding to $\mathcal{O}_{X_\beta}(s - s_i) \in \Gamma(U_i, \mathscr{P}ic^0(X_\beta/B))$ under the canonical identification $\mathscr{P}ic^0(X/B)) = \mathscr{X} = \mathscr{P}ic^0(X_\beta/B)$.

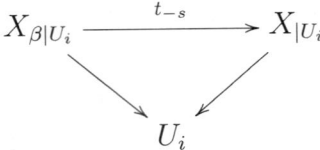

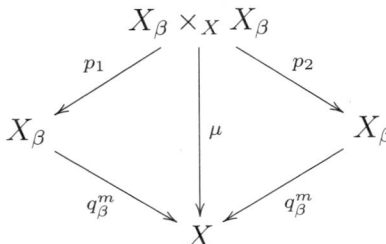

The formula (3.7) implies that

$$p_1^*\mathcal{O}_{X_\beta}(m(s-s_i)) \cong \mu^*\mathcal{O}_X(s-s_i)$$
$$p_2^*\mathcal{O}_{X_\beta}(m(s_i-s)) \cong \mu^*\mathcal{O}_X(s_i-s)$$

and so (3.6) holds. The theorem is proven. □

4. Duality between the lifting and extension presentations

We are now ready to prove Theorem B for a smooth elliptic fibration

$$X \underset{\sigma}{\overset{\pi}{\rightleftarrows}} B,$$

over a smooth space B satisfying $Br'_{an}(B) = 0$.

Let $\alpha, \beta, \varphi \in \text{Ш}_{an}(X)$ satisfy $m\beta = 0$, $m\varphi = \alpha$ (in particular α and β are m-compatible). We want to compare the derived categories of coherent sheaves on ${}_\alpha X_\beta$ and ${}_\beta X_\alpha$.

4.1. The gerby Fourier-Mukai transform. Let $D_1^b({}_\alpha X_\beta)$ and $D_{-1}^b({}_\beta X_\alpha)$ denote the derived categories of coherent sheaves of weight one and minus one on the gerbes ${}_\alpha X_\beta$ and ${}_\beta X_\alpha$ respectively. Alternatively, as explained at the end of Section 1.2, we can view $D_1^b({}_\alpha X_\beta)$ and $D_{-1}^b({}_\beta X_\alpha)$ as derived categories of $T_\beta(\alpha)$-twisted sheaves on X_β and $T_\alpha(-\beta)$ twisted sheaves on X_α respectively.

We want to construct a Fourier-Mukai functor

$$\boldsymbol{FM} : D_1^b({}_\alpha X_\beta) \to D_{-1}^b({}_\beta X_\alpha)$$

which is an equivalence. To achieve this we will work with the models ${}_\alpha\mathscr{L}_\beta$ and ${}_\beta\mathscr{E}_\alpha$ for ${}_\alpha X_\beta$ and ${}_\beta X_\alpha$ respectively. The idea is to use the explicit presentations for these models of the gerbes and construct the functor $\boldsymbol{FM}$ in terms of data on the atlases.

Since ${}_\alpha\mathscr{L}_\beta = [{}_\alpha LU_\beta / {}_\alpha LR_\beta]$ and ${}_\beta\mathscr{E}_\alpha = [{}_\beta EU_\alpha / {}_\beta ER_\alpha]$ we have natural structure morphisms

$$\gamma_L : \quad {}_\alpha LU_\beta \longrightarrow {}_\alpha\mathscr{L}_\beta$$

$$\gamma_E : \quad {}_\beta EU_\alpha \longrightarrow {}_\beta\mathscr{E}_\alpha$$

for the lifting and extension presentations. The (derived) pullback by γ_L gives a natural functor

$$\gamma_L^* : D_1^b({}_\alpha\mathscr{L}_\beta) \to D^b({}_\alpha LU_\beta),$$

which sends complexes of sheaves on ${}_\alpha\mathscr{L}_\beta$ to objects in $D^b({}_\alpha LU_\beta)$ preserved by the relations.

Explicitly, for a $\mathcal{L} \in D_1^b({}_\alpha\mathscr{L}_\beta)$, the pullback $\gamma_L^*\mathcal{L}$ is given by a pair $(L, \boldsymbol{f})$ where:

- L is a bounded complex of sheaves on the atlas ${}_{\alpha}LU_{\beta} = X_{\varphi} \times_B X_{\beta}$.
- $\boldsymbol{f} : p_{13}^* L \overset{q.i.}{\to} p_{23}^* L \otimes \mathscr{P}_{1-2, m\cdot 3}$ is a quasi-isomorphism of complexes on $X_{\varphi} \times_B X_{\varphi} \times_B X_{\beta}$, satisfying the cocycle condition (2.3).

Here p_{ij} is the projection of $X_{\varphi} \times_B X_{\varphi} \times_B X_{\beta}$ onto the product of the i-th and j-th components.

Under the gerby Fourier-Mukai transform, $\mathcal{L}$ should go to an object $\mathcal{Q} \in D^b_{-1}({}_{\beta}\mathscr{E}_{\alpha})$. To produce this object we will perform an integral transform from the derived category of the atlas ${}_{\alpha}LU_{\beta}$ to the derived category of the atlas ${}_{\beta}EU_{\alpha}$. Again, we would like to use the fact that the pullback by $\boldsymbol{\gamma}_E$ gives a functor

$$\boldsymbol{\gamma}_E^* : D^b_{-1}({}_{\beta}\mathscr{E}_{\alpha}) \to D^b({}_{\beta}EU_{\alpha}),$$

which sends complexes on ${}_{\beta}\mathscr{E}_{\alpha}$ to objects in $D^b({}_{\beta}EU_{\alpha})$ preserved by the relations. In principle this is all we can say about the images of $\boldsymbol{\gamma}_E^*$ since even for schemes the derived categories of coherent sheaves do not necessarily glue, see e.g. [**Har66**]. However in the case of ${}_{\beta}\mathscr{E}_{\alpha}$ we can be much more precise. It is known [**Pol96**, Theorem A] that given a scheme S and a finite flat morphism $p : U \to S$, the derived category of coherent sheaves on S is equivalent to the category of pairs $(F, \boldsymbol{\phi})$, where $F \in D^b(U)$ and $\boldsymbol{\phi} : p_1^* F \widetilde{\to} p_2^* F$ is an isomorphism in $D^b(U \times_S U)$ satisfying the cocycle condition in $D^b(U \times_S U \times_S U)$. If in addition we are given a $\mathbb{C}^{\times}$ bundle $R \to U \times_S U$ equipped with a biextension isomorphism, so that $[U/R]$ is a $\mathcal{O}^{\times}$-gerbe on S, we can repeat the reasoning of [**Pol96**, Theorem A] verbatim to conclude that the category $D^b_1([U/R])$ is equivalent to the category of pairs $(G, \boldsymbol{\psi})$, where $G \in D^b(U)$ and $\boldsymbol{\psi} : p_1^* G \widetilde{\to} p_2^* G \otimes R$ is an isomorphism in $D^b(U \times_S U)$ satisfying the cocycle condition (2.3) in $D^b(U \times_S U \times_S U)$. In particular, since by construction the morphism ${}_{\beta}EU_{\alpha} = X_{\varphi} \to X_{\alpha}$ is finite and flat, we conclude that $\boldsymbol{\gamma}_E^*$ identifies $D^b_{-1}({}_{\beta}\mathscr{E}_{\alpha})$ with the category of pairs $(Q, \boldsymbol{g})$ where:

- Q is a bounded complex of sheaves on the atlas ${}_{\beta}EU_{\alpha} = X_{\varphi}$.
- $\boldsymbol{g} : a_{\varphi}^* Q \overset{q.i.}{\to} p_2^* Q \otimes p_1^*(M_{\beta}^{-1} \otimes M_0)$ is a quasi-isomorphism of complexes on $X[m] \times_B X_{\varphi}$, satisfying the cocycle condition (2.3).

Here p_i is the projection of $X[m] \times_B X_{\varphi}$ onto the i-th component.

REMARK 3.3. The reader may wish to focus on the case when $\mathcal{L}$ is a line bundle on ${}_{\alpha}\mathscr{L}_{\beta}$ of fiber degree zero, i.e. $L \to X_{\varphi} \times_B X_{\beta}$ is a line bundle with $\deg(L_{|\{x\}\times_B X_\beta}) = 0$ and the existence of $\boldsymbol{f}$ is equivalent to having isomorphisms $L_{|X_\varphi \times_B \{y\}} \cong \mathcal{O}(my) \otimes \Sigma_\beta^{-1}$ for all $y \in X_\beta$.

In this case the object $\mathcal{Q}$ should be a spectral datum on the gerbe ${}_\beta\mathscr{E}_\alpha$ whose support is of degree one over B, i.e. Q is a torsion sheaf on X_φ supported on $(q_\varphi^n)^{-1}(s)$ for some section $s \subset X_\alpha$.

We will construct the functor $\boldsymbol{FM}$ by first constructing a functor between the derived categories on the atlases and then checking that this functor preserves the relations.

On the level of atlases, consider the functor

$$p_{1*} : D^b(X_\varphi \times_B X_\beta) \to D^b(X_\varphi).$$

We now have the following:

PROPOSITION 3.4. *The functor $p_{1*} : D^b(X_\varphi \times_B X_\beta) \to D^b(X_\varphi)$ preserves the relations defining ${}_\alpha\mathscr{L}_\beta$ and ${}_\beta\mathscr{E}_\alpha$ and so descends to a well defined exact functor*

$$\boldsymbol{FM} := (\boldsymbol{\gamma}_E^*)^{-1} \circ p_{1*} \circ \boldsymbol{\gamma}_L^* : D_1^b({}_\alpha X_\beta) \to D_{-1}^b({}_\beta X_\alpha).$$

Proof. Let $\boldsymbol{\gamma}_L^* \mathcal{L} = (L, \boldsymbol{f})$ be as above. Let $Q := p_{1*}L$. We need to construct a quasi-isomorphism of complexes on $X[m] \times_B X_\varphi$

$$\boldsymbol{g} : a_\varphi^* Q \to p_2^* Q \otimes p_1^*(M_\beta^{-1} \otimes M_0) \tag{3.8}$$

which depends functorially on $\boldsymbol{f}$.

We start by noting that there is a natural commutative diagram

$$\begin{array}{ccc}
X_\varphi \times_B X_\varphi \times_B X_\beta & \overset{p_{13}}{\underset{p_{23}}{\rightrightarrows}} & X_\varphi \times_B X_\beta \\
\uparrow & & \| \\
X_\varphi \times_{X_\alpha} X_\varphi \times_B X_\beta & \overset{\mathrm{pr}_{13}}{\underset{\mathrm{pr}_{23}}{\rightrightarrows}} & X_\varphi \times_B X_\beta \\
\downarrow{\scriptstyle \mathrm{pr}_{12}} & & \downarrow{\scriptstyle p_1} \\
X_\varphi \times_{X_\alpha} X_\varphi & \overset{\mathrm{pr}_{13}}{\underset{\mathrm{pr}_{23}}{\rightrightarrows}} & X_\varphi
\end{array}$$

Since the two bottom squares are fiber products we have the base change formulas:

$$\begin{aligned} \mathrm{pr}_1^* p_{1*}L &= p_{12*}\, \mathrm{pr}_{13}^* L \\ \mathrm{pr}_2^* p_{1*}L &= p_{12*}\, \mathrm{pr}_{23}^* L. \end{aligned} \tag{3.9}$$

Also, using the isomorphism $a_\varphi \times \mathrm{id} : X[m] \times_B X_\varphi \to X_\varphi \times_{X_\alpha} X_\varphi$ we reduce the problem of finding the map (3.8) to the equivalent problem of constructing a map

$$\mathrm{pr}_1^* p_{1*} L \to \mathrm{pr}_2^* p_{1*} L \otimes \Phi_\beta. \tag{3.10}$$

of complexes on $X_\varphi \times_{X_\alpha} X_\varphi$. Now using (3.9) and adjunction, this becomes

$$\mathrm{pr}_{13}^* L \to \mathrm{pr}_{23}^* L \otimes \mathrm{pr}_{12}^* \Phi_\beta, \tag{3.11}$$

on $X_\varphi \times_{X_\alpha} X_\varphi \times_B X_\beta$. Since $\mathrm{pr}_{13}^* L$ and $\mathrm{pr}_{23}^* L$ are the restrictions of $p_{13}^* L$ and $p_{23}^* L$, respectively, from $X_\varphi \times_B X_\varphi \times_B X_\beta$ to $X_\varphi \times_{X_\alpha} X_\varphi \times_B X_\beta$, the restriction of our map $\boldsymbol{f}$ gives

$$\boldsymbol{f} : \mathrm{pr}_{13}^* L \to \mathrm{pr}_{23}^* L \otimes (\mathscr{P}_{1\text{–}2,\, m\cdot 3 | X_\varphi \times_{X_\alpha} X_\varphi \times_B X_\beta}).$$

Therefore, in order to reconstruct from $\boldsymbol{f}$ a map (3.11) (equivalently $\boldsymbol{g}$), it suffices to exhibit a canonical isomorphism

$$\mathscr{P}_{1\text{–}2,\, m\cdot 3 | X_\varphi \times_{X_\alpha} X_\varphi \times_B X_\beta} \cong \mathrm{pr}_{12}^* \Phi_\beta, \tag{3.12}$$

on $X_\varphi \times_{X_\alpha} X_\varphi \times_B X_\beta$. As a first step in establishing (3.12) we note that both sides are pullbacks of sheaves on $X[m]$. On the right hand side, Φ_β was defined as $d^*(M_\beta^{-1} \otimes M_0)$ for the difference map $d : X_\varphi \times_{X_\alpha} X_\varphi \to X[m]$. On the left hand side, it suffices (in view of the see-saw principle) to argue that, for a point $\xi \in X[m]$, the restriction

$$\mathscr{P}_{1\text{–}2,\, m\cdot 3 | \{\xi\} \times_B X_\varphi \times_B X_\beta}$$

is trivial. But by the definition of $\mathscr{P}_{1\text{–}2,\, m\cdot 3}$ (see Section 1.1), this restriction can be identified with $\xi^{\otimes m}$. Since ξ has order m, we are done.

To conclude the construction of the map (3.12) and the proof of the proposition, we need to show that the line bundle $M_\beta^{-1} \otimes M_0$ on $X[m]$ is isomorphic to the direct image $R^0 p_{1-2*}(\mathscr{P}_{1\text{–}2,\, m\cdot 3 | X_\varphi \times_{X_\alpha} X_\varphi \times_B X_\beta})$. For this it is useful to identify $X_\varphi \times_{X_\alpha} X_\varphi \times_B X_\beta$ with $X[m] \times_B X_\varphi \times_B X_\beta$. Under that identification $\mathscr{P}_{1\text{–}2,\, m\cdot 3 | X_\varphi \times_{X_\alpha} X_\varphi \times_B X_\beta}$ becomes the pullback of $\mathscr{P}_{|X[m] \times_B X}$ under the natural map

$$\begin{array}{ccc} X[m] \times_B X_\varphi \times_B X_\beta & \longrightarrow & X[m] \times_B X \\ (\xi, x, y) & \longmapsto & (\xi, q_\beta(y)). \end{array}$$

Thus, the existence of the isomorphism (3.12) is equivalent to the existence of an isomorphism

$$\boldsymbol{p}_{1,m\cdot 2}^* (\mathscr{P}_{|X[m] \times_B X_\beta}) \cong \boldsymbol{p}_1^* (M_\beta^{-1} \otimes M_0), \tag{3.13}$$

where $\boldsymbol{p}_1 : X[m] \times_B X_\beta \to X[m]$ is the projection and $\boldsymbol{p}_{1,m\cdot 2} : X[m] \times_B X_\beta \to X[m] \times_B X$ is the map given by $(\xi, x) \mapsto (\xi, q_\beta(x))$.

Recall from Section 1.2 that by definition we have

$$\boldsymbol{p}_1^* M_\beta = a_\beta^* \Sigma_\beta \otimes \boldsymbol{p}_2^* \Sigma_\beta^{-1}.$$

Here $a_\beta : X[m] \times_B X_\beta$ is the action, $\boldsymbol{p}_2 : X[m] \times_B X_\beta \to X_\beta$ is the projection on the second factor and Σ_β is a line bundle on X_β of fiber degree m, which corresponds to the 'multiplication by m' map $q_\beta : X_\beta \to X$.

Look at the embedding $X[m] \times_B X_\beta \subset X \times_B X_\beta$. The projections $\boldsymbol{p}_1, \boldsymbol{p}_2$ and the maps $\boldsymbol{p}_{1,m\cdot 2}$ and a_β extend to the natural projections $X \times_B X_\beta \to X$ and $X \times_B X_\beta \to X_\beta$ and maps $X \times_B X_\beta \to X \times_B X$ and $X \times_B X_\beta \to X_\beta$, which we will denote by the same letters.

With this notation we have

LEMMA 3.5. *Let $\mathscr{P} \to X \times_B X$ denote the standard Poincare bundle. Then we have a natural isomorphism*

$$\boldsymbol{p}_{1,m\cdot 2}^* \mathscr{P} \otimes a_\beta^* \Sigma_\beta \otimes \boldsymbol{p}_2^* \Sigma_\beta^{-1} \cong \boldsymbol{p}_1^* \mathcal{O}_X(m\sigma).$$

Proof of the lemma. We will use the see-saw principle. Let $\xi \in X$ and let $b = \pi(\xi) \in B$. Then by viewing ξ as a line bundle of degree zero on $(X_\beta)_b$ and using the fact that $\Sigma_{\beta|X_b}$ is of degree m we compute

$$\begin{aligned} \boldsymbol{p}_{1,m\cdot 2}^* \mathscr{P}_{|\{\xi\}\times(X_\beta)_b} &= q_\beta^*(\xi) = \xi^{\otimes m}, \\ (a_\beta^* \Sigma_\beta \otimes \boldsymbol{p}_2^* \Sigma_\beta^{-1})_{|\{\xi\}\times(X_\beta)_b} &= t_\xi^* L_\beta \otimes L_\beta^{-1} = \xi^{\otimes -m}. \end{aligned}$$

Thus for every ξ we have

$$(\boldsymbol{p}_{1,m\cdot 2}^* \mathscr{P} \otimes a_\beta^* \Sigma_\beta \otimes \boldsymbol{p}_2^* \Sigma_\beta^{-1})_{|\{\xi\}\times(X_\beta)_b} \cong \mathcal{O}_{(X_\beta)_b} = \mathcal{O}_{(X_\beta)_b}$$

and so by the see-saw principle $D_\beta := R^0 \boldsymbol{p}_{1*}(\boldsymbol{p}_{1,m\cdot 2}^* \mathscr{P} \otimes a_\beta^* \Sigma_\beta \otimes \boldsymbol{p}_2^* \Sigma_\beta^{-1})$ is a line bundle on X satisfying

$$\boldsymbol{p}_{1,m\cdot 2}^* \mathscr{P} \otimes a_\beta^* \Sigma_\beta \otimes \boldsymbol{p}_2^* \Sigma_\beta^{-1} \cong \boldsymbol{p}_1^* D_\beta.$$

To compute the bundle D_β we consider a point $x \in X_\beta$. Let $b = \pi_\beta(x)$. Restricting to $X_b \times \{x\}$ we get

$$\begin{aligned} \boldsymbol{p}_{1,m\cdot 2}^* \mathscr{P}_{|X_b\times\{x\}} &= q_\beta(x) \quad \text{(viewed as degree zero line bundles on } X_b\text{)} \\ a_\beta^* \Sigma_{\beta|X_b\times\{x\}} &= t_x^* \Sigma_\beta \\ \boldsymbol{p}_2^* \Sigma_{\beta|X_b\times\{x\}} &= \mathcal{O}_{X_b}. \end{aligned}$$

Next, by the defining relationship between q_β and Σ_β we have that $q_\beta(x)$ is the line bundle of degree zero on X corresponding to $\mathcal{O}_{(X_\beta)_b}(mx) \otimes \Sigma_\beta^{-1}$ under the identification $\mathrm{Pic}^0((X_\beta)_b) = \mathrm{Pic}^0(X_b)$. Also, by the definition of a translation we have that $t_x^*\Sigma_\beta$ is the tensor product of $\mathcal{O}_{X_b}(m\sigma(b))$ with the line bundle of degree zero on X_b corresponding to $\mathcal{O}_{(X_\beta)_b}(-mx)\otimes\Sigma_\beta$ under the identification $\mathrm{Pic}^0((X_\beta)_b) = \mathrm{Pic}^0(X_b)$.

In other words, we have

$$(\boldsymbol{p}_{1,m\cdot 2}^*\mathscr{P}\otimes a_\beta^*\Sigma_\beta\otimes \boldsymbol{p}_2^*\Sigma_\beta^{-1})_{X_b\times\{x\}}\cong\mathcal{O}_{X_b}(m\sigma(b)),$$

for all $x\in X_\beta$. This implies that up to a twist by a pullback of a line bundle on B we have $D_\beta\cong\mathcal{O}_X(m\sigma)$. Finally, to fix the choice of this line bundle on B we look at the restriction of $\boldsymbol{p}_{1,m\cdot 2}^*\mathscr{P}\otimes a_\beta^*\Sigma_\beta\otimes\boldsymbol{p}_2^*\Sigma_\beta^{-1}$ on $\sigma\times_B X_\beta$ which is clearly isomorphic to $\mathcal{O}_{X_\beta}$. Hence the line bundle on B is trivial and the lemma is proven. □

In view of Lemma 3.5, the only thing left to check in order to establish the isomorphism (3.13) is that the line bundle M_0 on $X[m]$ is isomorphic to the restriction $\mathcal{O}_X(m\sigma)_{|X[m]}$. However applying the same reasoning we used in the proof of Lemma 3.5 to the projections $p_1,p_2 : X\times_B X\to X$ and the obvious maps $p_{1,m\cdot 2} : X\times_B X\to X\times_B X$ and $a_0 : X\times X\to X$, we see that $p_{1,m\cdot 2}^*\mathscr{P}\otimes a_0^*\mathcal{O}_X(m\sigma)\otimes p_2^*\mathcal{O}_X(-m\sigma)\cong p_1^*\mathcal{O}_X(m\sigma)$. On the other hand, from the definition of the Poincare bundle we have that $p_{1,m\cdot 2}^*\mathscr{P}_{|X[m]\times_B X}\cong\mathcal{O}$ and so $M_0\cong\mathcal{O}_X(m\sigma)_{|X[m]}$. This finishes the proof of the existence of (3.12). To complete the proof of the proposition, it only remains to note that since $(Q,\boldsymbol{g})$ was constructed from $(L,\boldsymbol{f})$ by means of the pushforward via $X_\varphi\times_B X_\beta\to X_\beta$ and the fixed isomorphism (3.12), it follows that $\boldsymbol{g}$ will satisfy the cocycle condition whenever $\boldsymbol{f}$ does. The proposition is proven. □

4.2. Categorical yoga for equivalences. We have constructed a functor $\boldsymbol{FM} : D_1^b({}_\alpha X_\beta)\to D_{-1}^b({}_\beta X_\alpha)$. We are going to prove that it is an equivalence. In order to do this, it is convenient to recall some general criteria, due to Bondal-Orlov and Bridgeland, for equivalences of triangulated categories.

Throughout this subsection we let $\boldsymbol{F} : \mathscr{A}\to\mathscr{B}$ be an exact functor between triangulated categories. A class Ω of objects in $\mathscr{A}$ is called a *spanning class* if for every $a\in\mathrm{ob}\,\mathscr{A}$, the left orthogonality condition

$$\mathrm{Hom}^i_{\mathscr{A}}(a,\omega)=0,\quad\text{for all}\quad i\in\mathbb{Z},\omega\in\Omega$$

implies that $a=0$, and similarly on the right. Recall the following

Theorem [BO95] *Assume that Ω is a spanning class for $\mathscr{A}$ and that the functor $\boldsymbol{F} : \mathscr{A}\to\mathscr{B}$ has left and right adjoints. Then $\boldsymbol{F}$ is fully*

faithful if and only if it is orthogonal:

$$\boldsymbol{F} : \mathrm{Hom}^i_{\mathscr{A}}(\omega_1, \omega_2) \widetilde{\to} \mathrm{Hom}^i_{\mathscr{B}}(\boldsymbol{F}\omega_1, \boldsymbol{F}\omega_2), \quad \textit{for all } i \in \mathbb{Z}, \omega_1, \omega_2 \in \Omega.$$

Assume now that our triangulated category $\mathscr{A}$ is linear. A functor $\boldsymbol{S}_{\mathscr{A}} : \mathscr{A} \to \mathscr{A}$ is called [**BK90**] a *Serre functor* for $\mathscr{A}$ if it is an exact equivalence and induces bifunctorial isomorphisms

$$\mathrm{Hom}_{\mathscr{A}}(a, b) \to \mathrm{Hom}_{\mathscr{A}}(b, \boldsymbol{S}_{\mathscr{A}} a)^{\vee}, \quad \text{for all } a, b \in \mathrm{ob}\, \mathscr{A},$$

satisfying compatibility with compositions. The basic example of a Serre functor is $\boldsymbol{S} : D^b(X) \to D^b(X)$, $\boldsymbol{S}a := a \otimes \omega_X[n]$, where X is a smooth n-dimensional projective variety and ω_X is the canonical bundle of X. If a Serre functor exists, it is unique up to an isomorphism of functors. We are now ready to state the main equivalence criterion we will be using:

Theorem [**Bri99, BKR01**] *Assume that $\mathscr{A}$ and $\mathscr{B}$ have Serre functors $\boldsymbol{S}_{\mathscr{A}}$, $\boldsymbol{S}_{\mathscr{B}}$, that $\mathscr{A} \neq 0$, $\mathscr{B}$ is indecomposable, and that $\boldsymbol{F} : \mathscr{A} \to \mathscr{B}$ has a left adjoint. Then $\boldsymbol{F}$ is an equivalence if it is fully faithful and it intertwines the Serre functors: $\boldsymbol{F} \circ \boldsymbol{S}_{\mathscr{A}}(\omega) = \boldsymbol{S}_{\mathscr{B}} \circ \boldsymbol{F}(\omega)$ on all elements $\omega \in \Omega$ in the spanning class.*

We want to show that our gerby Fourier-Mukai functor

$$\boldsymbol{FM} : D^b_1({}_{\alpha}X_{\beta}) \to D^b_1({}_{\beta}X_{\alpha})$$

is an equivalence. The results above suggest that we should first exhibit Serre functors for $D^b_1({}_{\alpha}X_{\beta})$ and $D^b_1({}_{\beta}X_{\alpha})$, and find a suitable spanning class for $D^b_1({}_{\alpha}X_{\beta})$. These results, which do not involve $\boldsymbol{FM}$, are carried out in section 4.3 below. In section 4.4 we then complete the argument by showing that our $\boldsymbol{FM}$ preserves orthogonality and intertwines the Serre functors.

4.3. Serre functors and spanning classes for $\mathcal{O}^{\times}_X$-gerbes. Let X be an n-dimensional smooth projective variety. Let $c : {}_{\alpha}X \to X$ be an $\mathcal{O}^{\times}_X$-gerbe corresponding to an element $\alpha \in H^2(X, \mathcal{O}^{\times}_X)$.

CLAIM 3.6. *The functor*

$$\begin{array}{rccc} \boldsymbol{S} : & D^b_1({}_{\alpha}X) & \longrightarrow & D^b_1({}_{\alpha}X) \\ & a & \longmapsto & a \otimes c^* \omega_X[n] \end{array}$$

is a Serre functor.

Proof. For any $a, b \in D_1^b({}_\alpha X)$, we need a natural isomorphism

$$\mathrm{Hom}_{D_1^b({}_\alpha X)}(a,b) \to \mathrm{Hom}_{D_1^b({}_\alpha X)}(b, \boldsymbol{S}a)^\vee.$$

Since a, b have weight 1, $\mathscr{R}Hom_{{}_\alpha X}(a,b)$ has weight 0, so there exists a unique $\mathscr{H}(a,b) \in D^b(X)$ such that

$$\mathscr{R}Hom_{{}_\alpha X}(a,b) = c^*\mathscr{H}(a,b).$$

It follows that

$$\mathrm{Hom}_{D_1^b({}_\alpha X)}(a,b) = R\Gamma_X(\mathscr{H}(a,b)).$$

Similarly,

$$\mathscr{R}Hom_{{}_\alpha X}(b, \boldsymbol{S}a) = c^*(\mathscr{H}(b,a) \otimes \omega_X[n]),$$

so

$$\mathrm{Hom}_{D_1^b({}_\alpha X)}(b, \boldsymbol{S}a)^\vee = R\Gamma_X(\mathscr{H}(b,a) \otimes \omega_X[n])^\vee = R\Gamma_X(\mathscr{H}(b,a)^\vee),$$

where the last step uses the usual Serre duality. So all we need is the identification

$$\mathscr{H}(a,b) \tilde{\rightarrow} \mathscr{H}(b,a)^\vee,$$

or a non-degenerate pairing on $\mathscr{H}(a,b) \times \mathscr{H}(b,a)$. But since

$$c^* : D^b(X) \tilde{\rightarrow} D_0^b({}_\alpha X)$$

is an equivalence of categories, this follows immediately from the non-degenerate pairing on $\mathscr{R}Hom_{{}_\alpha X}(a,b) \times \mathscr{R}Hom_{{}_\alpha X}(b,a)$ given by the trace. □

Since our functor $\boldsymbol{F} = \boldsymbol{FM}$ was constructed as a push-forward on the atlases, it has an obvious left adjoint $\boldsymbol{G}$ corresponding to the pullback functor on the atlases. Therefore $\boldsymbol{FM}$ also has a right adjoint, namely $S_{D_1^b({}_\alpha X_\beta)} \circ \boldsymbol{G} \circ S^{-1}_{D_1^b({}_\beta X_\alpha)}$.

Fix a point $p \in X_\beta$. Since the restriction of ${}_\alpha X_\beta$ to p is the trivial gerbe on p for any α, the torsion sheaf $\mathcal{O}_p$ can be considered as a weight one sheaf on ${}_\alpha X_\beta$ for any α. For our spanning class Ω we take the structure sheaves of points on the space X_β, viewed as sheaves of weight one on the stack ${}_\alpha X_\beta$.

CLAIM 3.7. *Let $c : {}_\alpha X \to X$ be an $\mathcal{O}_X^\times$-gerbe on a smooth projective X. Then the class Ω consisting of structure sheaves $\mathcal{O}_p$ of points on X, viewed as sheaves of weight one on ${}_\alpha X$, is a spanning class for $D_1^b({}_\alpha X)$.*

Proof. In order to show that the class Ω is a spanning class, we need to show that for every $a \in \operatorname{ob} D_1^b({}_\alpha X)$, the left orthogonality condition

$$\operatorname{Hom}^i_{D_1^b({}_\alpha X)}(a, \mathcal{O}_p) = 0, \quad \text{for all} \quad i \in \mathbb{Z}, p \in X$$

implies that $a = 0$. We also need the analogous result on the right, but this follows using the Serre functor. We can also reduce to the case that a is represented by a single sheaf on ${}_\alpha X$, i.e. a is an α-twisted sheaf on X. Now such an a is specified in terms of its sections on an appropriate etale atlas U plus some α-twisted gluing. In order to conclude that $a = 0$, it suffices to show that every $p \in X$ has a neighborhood U' on which $a = 0$. But by restricting to a small enough neighborhood U' of p in U, we can get the class α to vanish. The restriction of a to U' and the $\mathcal{O}_p$ for $p \in U'$ become ordinary sheaves. The group $\operatorname{Hom}^i_{D_1^b({}_\alpha X)}(a, \mathcal{O}_p)$ can be computed on either U or U'. Therefore, the orthogonality condition forces a to vanish on U', which is what we need. □

4.4. Orthogonality and intertwining. Now that we have a Serre functor and a spanning class, we are ready to apply the general results of subsection 4.2 to our gerby Fourier-Mukai functor $\boldsymbol{FM}$.

CLAIM 3.8. *The gerby Fourier-Mukai functor*

$$\boldsymbol{FM} : D_1^b({}_\alpha X_\beta) \to D_{-1}^b({}_\beta X_\alpha)$$

is orthogonal on Ω*:*

$$\boldsymbol{FM} : \operatorname{Hom}^i_{D_1^b({}_\alpha X_\beta)}(\mathcal{O}_{x_1}, \mathcal{O}_{x_2}) \widetilde{\to} \operatorname{Hom}^i_{D_1^b({}_\beta X_\alpha)}(\boldsymbol{F}\mathcal{O}_{x_1}, \boldsymbol{F}\mathcal{O}_{x_2}),$$

for all $i \in \mathbb{Z}, x_1, x_2 \in X_\beta$.

Proof. Recall (3.4) that our functor $\boldsymbol{FM}$ descends from $p_{1*} : D^b(X_\varphi \times_B X_\beta) \to D^b(X_\varphi)$. Let $b_1, b_2 \in B$ be the images of $x_1, x_2 \in X_\beta$. If $b_1 \neq b_2$ then the supports are disjoint, so the Hom^i on both sides clearly vanish. Assume then that $b_1 = b_2 = b$. In this case, the structure sheaf $\mathcal{O}_{{}_\alpha \mathscr{L}_\beta | x_i}$ is supported on the fiber $C_{x_i} = X_\varphi \times_B (x_i)$, and both supports map isomorphically to $C_b = X_\varphi \times_B (b) \subset X_\varphi$. Now $\boldsymbol{FM}(\mathcal{O}_{{}_\alpha \mathscr{L}_\beta | x_i})$ is the line bundle $\mathscr{L}_\beta(-mx_i)$ on C_b, so

$$\begin{aligned} \operatorname{Hom}^i_{D_1^b({}_\beta X_\alpha)}(\boldsymbol{FM}\mathcal{O}_{x_1}, \boldsymbol{FM}\mathcal{O}_{x_2}) &= \operatorname{Hom}_{C_b}(\mathscr{L}_\beta(-mx_1), \mathscr{L}_\beta(-mx_2)) \\ &= \operatorname{Hom}^i_{D_1^b({}_\alpha X_\beta)}(\mathcal{O}_{x_1}, \mathcal{O}_{x_2}), \end{aligned}$$

completing the proof. □

We note that both sides of the claim vanish unless x_1, x_2 differ by a point of $X[m]$, in which case they define isomorphic sheaves. The spanning class Ω may therefore be taken to be parametrized by $X = X_\beta/X[m]$ rather than by X_β.

CLAIM 3.9. *The gerby Fourier-Mukai functor* $\boldsymbol{FM} : D^b_1({}_\alpha X_\beta) \to D^b_{-1}({}_\beta X_\alpha)$ *intertwines the Serre functors, i.e.:* $\boldsymbol{FM} \circ \boldsymbol{S}_{{}_\alpha X_\beta}(\mathcal{O}_p) = \boldsymbol{S}_{{}_\beta X_\alpha} \circ \boldsymbol{FM}(\mathcal{O}_p)$ *for all points* $p \in X_\beta$.

Proof. Follows immediately from the fact that $\boldsymbol{FM}\mathcal{O}_x$ is supported on C_b and that the canonical sheaf of X_β restricts to the trivial line bundle on C_b. □

CHAPTER 4

Surfaces

In case X is a surface, we can refine the previous results to include the singular fibers. On a surface, any pair of classes $\alpha, \beta \in \text{III}_{an}(X)$ are complementary in the sense of subsection 3, by Corollary 2.17, so the gerbes ${}_{\alpha}X_{\beta}, {}_{\beta}X_{\alpha}$ are always well- defined. When $m\beta = 0$ we will construct the lifting presentation of ${}_{\alpha}X_{\beta}$ and the extension presentation of ${}_{\beta}X_{\alpha}$. Then we will exhibit a Fourier-Mukai transform $\boldsymbol{FM}$ between these presentations. Finally, we will show that $\boldsymbol{FM}$ is an equivalence of categories by verifying the criterion of Bondal-Orlov and Bridgeland.

We assume throughout that X is a smooth surface, B is a smooth curve, and the elliptic fibration $\pi : X \to B$ has at most singular fibers of type I_1. Since every such elliptic surface is uniquely determined by its monodromy representation it is clear that we can always extend X to a smooth compact relatively minimal elliptic surface whose base curve is a suitable compactification of B. Furthermore, by Kodaira's classification of compact complex surfaces it follows that every smooth compact elliptic surface is Kähler (in fact algebraic). Therefore X must be Kähler as well.

1. The lifting presentation

Our first goal is to construct the lifting presentation of ${}_{\alpha}X_{\beta}$, in a way that restricts to the previously constructed presentation on the non-singular fibers. We start with the second projection $p_2 : X_{\varphi} \times_B X_{\beta} \to X_{\beta}$. Unfortunately, this is *not* an atlas for the gerbe ${}_{\alpha}X_{\beta}$. The problem can be traced back to the fact that the threefold $X_{\varphi} \times_B X_{\beta}$ is singular. So let

$$Y := \widehat{X_{\varphi} \underset{B}{\times} X_{\beta}}$$

be a small resolution of $X_{\varphi} \times_B X_{\beta}$. Now Y is smooth and equipped with flat morphisms $\nu_1 : Y \to X_{\varphi}$ and $\nu_2 : Y \to X_{\beta}$ which lift p_1 and p_2 respectively. There is an induced map

$$\nu_2^* : Br'_{an}(X_{\beta}) \to Br'_{an}(Y).$$

We claim that Y is an atlas for ${}_{\alpha}X_{\beta}$, i.e. that ${}_{\alpha}X_{\beta}$ has a presentation:

$$
(4.1)\qquad
\begin{array}{ccccc}
{}_\alpha LR_\beta & & & & \\
\downarrow & \overset{s}{\underset{t}{\searrow\searrow}} & & & \\
{}_\alpha LU_\beta \times_{X_\beta} {}_\alpha LU_\beta & \overset{p_1}{\underset{p_2}{\rightrightarrows}} & {}_\alpha LU_\beta & \xrightarrow{\nu_2} & X_\beta
\end{array}
$$

where

$$
\begin{aligned}
{}_\alpha LU_\beta &:= Y, \\
\nu_2 : Y \to X_\beta &\quad \text{is the second projection, and} \\
{}_\alpha LR_\beta &:= Y \times_{{}_\alpha X_\beta} Y.
\end{aligned}
$$

We call (4.1) the *Lifting Presentation* of ${}_\alpha X_\beta$. By Remark 2.4 (iv), the fact that (4.1) is indeed a presentation follows from:

LEMMA 4.1. $\nu_2^*(T_\beta(\alpha)) = 0$.

Proof. Let $B^o \subset B$ be the complement of the discriminant, and let $Y^o \subset Y, X^o \subset X$ be the inverse images of B^o. The maps

$$Y^o \overset{i}{\hookrightarrow} Y \overset{\nu_2}{\longrightarrow} X_\beta$$

lead via the exponential sequence to the diagram:

$$
\begin{array}{ccccccc}
 & & & & H^2_{an}(X_\beta, \mathcal{O}^\times) & \overset{\partial}{\to} & H^3(X_\beta, \mathbb{Z}) = 0 \\
 & & & & \downarrow{\scriptstyle \nu_2^*} & & \downarrow{\scriptstyle \nu_2^*} \\
0 & \longrightarrow & H^2_{an}(Y, \mathcal{O}_Y)/H^2(Y, \mathbb{Z}) & \overset{\exp_Y}{\hookrightarrow} & H^2_{an}(Y, \mathcal{O}^\times) & \overset{\partial}{\longrightarrow} & H^3(Y, \mathbb{Z}) \\
 & & \downarrow{\scriptstyle i^*_{\mathcal{O}}} & & \downarrow{\scriptstyle i^*_{\mathcal{O}^\times}} & & \downarrow \\
0 & \to & H^2_{an}(Y^o, \mathcal{O}_{Y^o})/H^2(Y^o, \mathbb{Z}) & \overset{\exp_{Y^o}}{\hookrightarrow} & H^2_{an}(Y^o, \mathcal{O}^\times) & \longrightarrow & H^3(Y^o, \mathbb{Z}).
\end{array}
$$

We have

$$\partial \nu_2^*(T_\beta(\alpha)) = \nu_2^* \partial(T_\beta(\alpha)) = \nu_2^* 0 = 0,$$

so

$$\nu_2^*(T_\beta(\alpha)) = \exp_Y(a)$$

for some $a \in H^2_{an}(Y, \mathcal{O}_Y)/H^2(Y, \mathbb{Z})$. We know from 3.1 that Y^o is an atlas for the restriction of $T_\beta(\alpha)$ to X^o, so

$$0 = i^*_{\mathcal{O}^\times} \nu_2^*(T_\beta(\alpha)) = i^*_{\mathcal{O}^\times} \exp_Y(a) = \exp_{Y^o} i^*_{\mathcal{O}}(a).$$

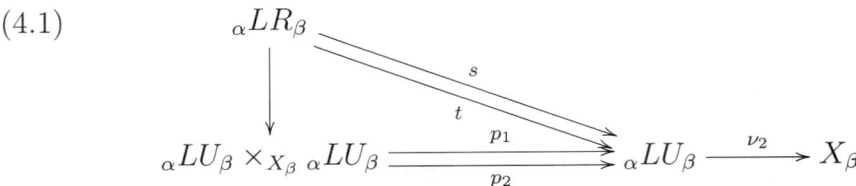

But $H^2_{an}(Y, \mathcal{O}_Y)/H^2(Y, \mathbb{Z})$ is a birational invariant, so $i^*_{\mathcal{O}}$ is an isomorphism. Since $\exp_{Y^o}$ is injective, we see that $a = 0$, so we are done. $\square$

COUNTER EXAMPLE 4.2. Let $A^o \subset A$ be an open subset in a smooth variety. By the birational invariance of cohomological Brauer groups [**Mil80**], the restriction map:

$$H^2_{et}(A, \mathcal{O}^\times_A) \to H^2_{et}(A^o, \mathcal{O}^\times_{A^o})$$

is injective. Nevertheless, the analogous map:

$$H^2_{an}(A, \mathcal{O}^\times_A) \to H^2_{an}(A^o, \mathcal{O}^\times_{A^o})$$

may fail to be injective. In our situation, all we were able to prove, and fortunately all that was needed, was that $H^2_{an}(Y, \mathcal{O}^\times_Y) \to H^2_{an}(Y^o, \mathcal{O}^\times_{Y^o})$ is injective on the image of $H^2_{an}(X_\beta, \mathcal{O}^\times_{X_\beta})$.

As an example, let $C \subset \mathbb{P}^3$ be a smooth curve of genus ≥ 2, let A be the blowup of $\mathbb{P}^3$ along C, and let $A^o := \mathbb{P}^3 \setminus C$. Then by the exponential sequence,

$$H^2_{an}(A^o, \mathcal{O}^\times_{A^o}) = H^3(A^o; \mathbb{Z}),$$
$$H^2_{an}(A, \mathcal{O}^\times_A) = H^3(A; \mathbb{Z}),$$

but by excision, $H^3(A^o; \mathbb{Z}) \cong H^3(\mathbb{P}^3, C; \mathbb{Z}) \cong \mathbb{Z}$, while $H^3(A; \mathbb{Z}) \cong H^1(C, \mathbb{Z})$.

2. The extension presentation

Next, we want to construct the extension presentation of ${}_\beta X_\alpha$, in a way that restricts to the previously constructed extension presentation on the complement of the singular fibers. Fix $\varphi \in Ш_{an}(X)$ satisfying $m \cdot \varphi = \alpha$. Let X^o_α, X^o_φ be the inverse images in X_α, X_φ of B^o, the complement of the discriminant in B. In the non-singular case, our atlas was given by the multiplication-by-m map $q_\varphi : X^o_\varphi \to X^o_\alpha$. Unfortunately, this does not extend to a morphism $q_\varphi : X_\varphi \to X_\alpha$. Instead, we will construct another (singular!) surface $\widehat{X}_\varphi$ with a birational morphism $\widehat{X}_\varphi \to X_\varphi$ and a flat morphism $\hat{q}_\varphi : \widehat{X}_\varphi \to X_\alpha$ which restricts to the previous q^o_φ. This data gives a commutative diagram:

$$\begin{array}{ccc} H^2_{an}(X_\alpha, \mathcal{O}^\times) & \xrightarrow{\hat{q}^*_\varphi} & H^2_{an}(\widehat{X}_\varphi, \mathcal{O}^\times) \\ \downarrow \cong & & \downarrow \cong \\ H^2_{an}(X^o_\alpha, \mathcal{O}^\times) & \xrightarrow{(q^o_\varphi)^*} & H^2_{an}(X^o_\varphi, \mathcal{O}^\times) \end{array}$$

The exponential sequence shows that the two vertical maps are isomorphisms, as in the proof of Lemma 4.1: this uses the fact that $H^2_{an}(\widehat{X}_\varphi, \mathcal{O}_{\widehat{X}_\varphi})/H^2(\widehat{X}_\varphi, \mathbb{Z})$ is a birational invariant, and that $\ker[H^3(\widehat{X}_\varphi, \mathbb{Z}) \to H^3(\widehat{X}_\varphi, \mathbb{R})] = 0$, $\ker[H^3(X_\alpha, \mathbb{Z}) \to H^3(X_\alpha, \mathbb{R})] = 0$, which in turn follows from the observation that the third cohomology of a smooth 4-manifold has no torsion and that $\widehat{X}_\varphi$ and X_α are Kähler surfaces. Since $(q^o_\varphi)^*$ kills all classes of order m, it follows that so does $\hat{q}^*_\varphi$, so $\widehat{X}_\varphi$ is indeed an atlas.

In order to construct $\widehat{X}_\varphi$ we have to resolve the rational map $q_\varphi : X_\varphi \dashrightarrow X_\alpha$. For that we can work locally in the complex topology on B near a point $p \in B \setminus B^o$, i.e. we can replace B by a small disc centered at p. Over this disc the group scheme $X^\sharp[m]$ has a subgroup scheme $I \subset X^\sharp[m]$ of cycles invariant under the local monodromy around p. Since by assumption the singular fibers of X are of type I_1, the group scheme I is isomorphic to $B \times (\mathbb{Z}/m)$ and consists of all the sections in $X[m]$ over the disc that pass through smooth points of the fiber X_p. Translations by such sections give rise to a well defined action of I on X_φ which fixes the singular point x_p of the fiber $(X_\varphi)_p$. Therefore over our disk the rational map q_φ decomposes as

$$\begin{array}{ccc} X_\varphi & \overset{q_\varphi}{\dashrightarrow} & X_\alpha \\ & {\scriptstyle s}\searrow \quad \nearrow & \\ & X_\varphi/I & \end{array}$$

where s denotes the quotient map. The surface X_φ/I has a unique singularity at the image of the point x_p and the map $X_\varphi/I \to B$ is a flat genus one fibration. A straightforward local computation at the singular point of the I_1 fiber of X_φ shows that in suitably chosen local coordinates (z, w) near x_p the generator of I acts as $(z, w) \mapsto (\zeta z, \zeta w)$, where ζ is a primitive m-th root of unity. This implies that the singularity of X_φ/I is of type A_{m-1}. The minimal resolution $\widehat{X_\varphi/I} \to X_\varphi/I$ of X_φ/I is a flat genus one fibration over B with a single I_m fiber over p.

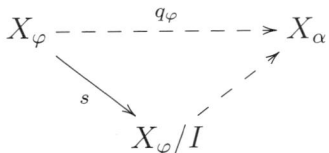

On the other hand, over our small disk, X_α retracts to the singular fiber $(X_\alpha)_p$ whose fundamental group is $\mathbb{Z}$. Therefore there is a unique m-sheeted etale cover $\widetilde{X}_\alpha \to X_\alpha$. By construction the covering map commutes with the projections $\tilde{\pi}_\alpha : \widetilde{X}_\alpha \to B$ and $\pi_\alpha : X_\alpha \to B$ and so the fiber $(\widetilde{X}_\alpha)_p$ of $\tilde{\pi}_\alpha$ over the point $p \in B$ is a Kodaira fiber of type I_m. The surfaces $\widetilde{X}_\alpha$ and $\widehat{X_\varphi/I}$ are clearly isomorphic outside of the preimages of $p \in B$ and so are birational. Since as genus one fibrations $\widetilde{X}_\alpha$ and $\widehat{X_\varphi/I}$ are both relatively minimal, this implies that $\widetilde{X}_\alpha$ and $\widehat{X_\varphi/I}$ are actually isomorphic due to the uniqueness [**Kod63**] of the relatively minimal models. Recall next that by the work of Ito and Nakamura [**IN99, BKR01**] the minimal resolution $\widehat{X_\varphi/I}$ of X_φ/I can be identified with the Hilbert scheme of I-clusters in X_φ. In the spirit of [**BKR01**] consider the universal closed subscheme $Z \subset X_\varphi \times \widehat{X_\varphi/I}$ with its natural projection to X_φ and $\widehat{X_\varphi/I} \cong \widetilde{X}_\alpha$. There is a commutative diagram of spaces

$$\begin{array}{ccccc} & & Z & & \\ & \swarrow & & \searrow & \\ \widetilde{X}_\alpha & & & & X_\varphi \\ & \searrow & & \swarrow & \\ & & X_\varphi/I & & \end{array}$$

where the morphism $Z \to X_\varphi$ is birational and surjective and the morphism $Z \to \widetilde{X}_\alpha$ is finite and flat. In particular the composite map

$$Z \to \widetilde{X}_\alpha \to X_\alpha \tag{4.2}$$

is a finite and flat morphism which extends the multiplication-by-m map $q^o_\varphi : X^o_\varphi \to X^o_\alpha$. It is also helpful to observe that the two intermediate maps appearing in the construction of (4.2) are both Galois covers with Galois group isomorphic to $\mathbb{Z}/m$. The first one is just the quotient of $X_\varphi \times_{X_\varphi/I} \widetilde{X}_\alpha$ by I and the second is the etale Galois cover $\widetilde{X}_\alpha \to X_\alpha$.

Our extension atlas $\widehat{X}_\varphi$ is obtained by gluing the local surfaces Z defined over small discs centered at discriminant points $p \in B \setminus B^o$ to X^o_φ. We will write $\varepsilon : \widehat{X}_\varphi \to X_\varphi$ for the contraction map and $\hat{q}_\varphi : \widehat{X}_\varphi \to X_\alpha$ for the finite flat map gluing each (4.2) to q^o_φ. Note that by construction the surface $\widehat{X}_\varphi$ is singular. It has isolated toroidal singularities sitting over the singular points of the singular fibers of $\widetilde{X}_\alpha$. In particular $\widehat{X}_\varphi$ is a normal analytic surface.

3. Duality for gerby genus one fibered surfaces

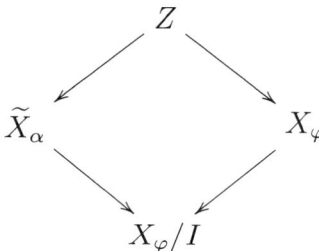

With the description of the global lifting and extension presentations for gerby genus one fibered surfaces in place, we are now ready to construct the Fourier-Mukai functor between the derived categories of pure weight one and to show that it is an equivalence. The key property of our gerby surfaces, which makes the construction possible is the fact that the gerbes appearing in the picture become trivial when we restrict our attention to a piece of the surface sitting over a sufficiently small open disk in B.

We recall our convention that all our direct and inverse images, as well as the tensor product, are taken in the derived category. Thus for any space Z, we will simply write $\otimes$ for the derived tensor product $\otimes^L$ on $D^b(Z)$ and for any map of spaces $p : Z \to T$ we will write $p_*, p^*, p_!, p^!$, for the corresponding derived functors (whenever these functors make sense on the bounded derived categories).

Following the pattern of the proof in section 4.1 we will construct a Fourier-Mukai functor

$$\boldsymbol{FM} : D^b_1({}_\alpha X_\beta) \to D^b_{-1}({}_\beta X_\alpha)$$

by exhibiting an integral transform between the derived categories on the atlases of the presentations ${}_\alpha\mathscr{L}_\beta$ and ${}_\beta\mathscr{E}_\alpha$ and then checking that this functor preserves the relations.

To avoid cumbersome notation we will write

$$Y := \widehat{X_\varphi \underset{B}{\times} X_\beta} \quad \text{and} \quad S := \widehat{X}_\varphi$$

for the atlases of the presentations ${}_\alpha\mathscr{L}_\beta$ and ${}_\beta\mathscr{E}_\alpha$ respectively.

With this notation the relations of ${}_\alpha\mathscr{L}_\beta$ are given by the total space

$$\operatorname{tot}(\mathscr{P}^{\times}_{1\text{–}2,\, m\cdot 3}) \longrightarrow Y \times_{X_\beta} Y \rightrightarrows Y.$$

Here $\mathscr{P}_{1\text{–}2,\, m\cdot 3}$ denotes an appropriate line bundle on $Y \times_{X_\beta} Y$ which extends the line bundle $p^*_{1-2,m3}\mathscr{P} \to X^o_\varphi \times_{B^o} X^o_\beta$ discussed in section 1.1. Note that we know that such a line bundle exists due to Lemma 4.1. Explicitly, the total space $\operatorname{tot}(\mathscr{P}^{\times}_{1-2,m3})$ is isomorphic to the stacky fiber product $Y \times_{{}_\alpha X_\beta} Y$ (this product is a space again by Lemma 4.1). We will write $Y^o = X^o_\varphi \times_{B^o} X^o_\beta$ for the part of Y sitting over B^o, and we put $\mathscr{P}^o_{1\text{–}2,\, m\cdot 3} := \mathscr{P}_{1\text{–}2,\, m\cdot 3|Y^o} = p^*_{1\text{–}2,\, m\cdot 3}\mathscr{P}$.

Similarly the relations of the presentation ${}_\beta\mathscr{E}_\alpha$ are given by the total space

$$\operatorname{tot}(\Phi^{\times}_\beta) \longrightarrow S \times_{X_\alpha} S \rightrightarrows S.$$

Here Φ_β denotes an appropriate line bundle which extends the line bundle $d^*(M_\beta \otimes M_0^{-1}) \to X^o_\varphi \times_{X^o_\alpha} X^o_\varphi$ discussed in section 1.2. As before, the existence of the line bundle Φ_β is guaranteed by the observation that the class $T_\alpha(\beta)$ of the extension gerbe vanishes (see section 2) when we pull it back to S. We will write $S^o = X^o_\varphi$ for the part of S sitting over B^o and we put $\Phi^o_\beta := \Phi_{\beta|S^o} = d^*(M_\beta \otimes M_0^{-1})$.

Using the above setup we can now identify the category $D^b_1({}_\alpha X_\beta)$ with the category of objects $L \in D^b(Y)$ equipped with descent datum $\boldsymbol{f} : p_1^* L \xrightarrow{q.i.} p_2^* L \otimes \mathscr{P}_{1\text{-}2, m\cdot 3}$ on $Y \times_{X_\beta} Y$, satisfying the cocycle condition described at the end of section 1.2. Similarly we identify $D^b_{-1}({}_\beta X_\alpha)$ with the category of objects $Q \in D^b(S)$ equipped with descent datum $\boldsymbol{g} : p_1^* Q \xrightarrow{q.i.} p_2^* Q \otimes \Phi_\beta^{-1}$ on $S \times_{X_\alpha} S$ satisfying the same cocycle condition as in section 1.2. These identifications reduce the problem of constructing the Fourier-Mukai functor $\boldsymbol{FM} : D^b_1({}_\alpha X_\beta) \to D^b_{-1}({}_\beta X_\alpha)$ to the problem of constructing a functor $\boldsymbol{F} : D^b(Y) \to D^b(S)$ which maps descent data to descent data.

We will define the functor $\boldsymbol{F}$ as an integral transform with respect to a suitable kernel object $\Pi \in D^b(Y \times S)$. We proceed to construct Π by gluing together certain locally defined coherent sheaves on $Y \times S$. We carry out this gluing in the analytic topology to obtain the general functor $\boldsymbol{F}$ we need. Note that in the algebraic case, the kernel Π still produces the correct functor Π in view of the GAGA principle.

First we look at the smooth part of the genus one fibrations X_α, X_β, X_φ, etc. As usual, we write $B^o \subset B$ for the complement of the discriminant of the map $\pi : X \to B$. Similarly, for any space (or stack) $Z \to B$ mapping to B, we put $Z^o := Z \times_B B^o$. The atlases Y^o and S^o for the gerbes ${}_\alpha\mathscr{L}_\beta{}^o$ and ${}_\beta\mathscr{E}_\alpha{}^o$ can be described simply as

$$\begin{aligned} Y^o &= X^o_\varphi \times_{B^o} X^o_\beta \\ S^o &= X^o_\varphi \end{aligned}$$

and so over B^o we recover the setup analyzed in section 4. In this setup the integral transform we need was defined as the pushforward $p_{1*} : D^b(X^o_\varphi \times_{B^o} X^o_\beta) \to D^b(X^o_\varphi)$ with respect to the projection on the first factor. Equivalently, we can view this functor as the integral transform whose kernel is the sheaf $\Delta_{\varphi *}\mathcal{O}_{X^o_\varphi \times_{B^o} X^o_\beta}$ on $X^o_\varphi \times_{B^o} X^o_\varphi \times_{B^o} X^o_\beta$, where $\Delta_\varphi(x, y) := (x, x, y)$ for all $(x, y) \in X^o_\varphi \times_{B^o} X^o_\beta$. In view of this we set:

$$\Pi^o := \mathcal{O}_{X^o_\varphi \times_{B^o} X^o_\beta} \in \mathrm{Coh}(X^o_\varphi \times_{B^o} X^o_\beta) = \mathrm{Coh}(Y^o \times_{X^o_\varphi} S^o).$$

Let now $p \in B \setminus B^o$. Choose small analytic discs $p \in U^p \subset B$ around each such p so that $U^p \cap U^q = \varnothing$ for $p \neq q$ and the genus one fibrations X_φ and X_β both admit analytic sections over each U^p. For any space (or stack) $Z \to B$ mapping to B we will write Z^p for the restriction $Z \times_B U^p$. Note that

$$Y \times_{X_\varphi} S = (Y^o \times_{X_\varphi^o} S^o) \bigcup \left(\coprod_{p \in B \setminus B^o} (Y^p \times_{X_\varphi^p} S^p) \right),$$

and so, in order to extend Π^o to a globally defined sheaf on $Y \times_{X_\varphi} S$, it suffices to construct coherent analytic sheaves Π^p on each $Y^p \times_{X_\varphi^p} S^p$ and isomorphisms

$$\Pi^o_{|(Y^p \times_{X_\varphi^p} S^p) \cap (Y^o \times_{X_\varphi^o} S^o)} \cong \Pi^p_{|(Y^p \times_{X_\varphi^p} S^p) \cap (Y^o \times_{X_\varphi^o} S^o)}.$$

Fix a $p \in B \setminus B^o$ and let $s_\varphi : U^p \to X_\varphi^p$ and $s_\beta : U^p \to X_\beta^p$ be local holomorphic sections. Since the rational map $q_\varphi : X_\varphi \dashrightarrow X_\alpha$ is a well defined morphism when restricted to the complement $X_\varphi^\sharp$ of the singular points of the singular fibers of $\pi_\varphi : X_\varphi \to B$, and since $s_\varphi(U^p) \subset X_\varphi^\sharp$, it follows that s_φ induces also a well defined section $s_\alpha := q_\varphi \circ s_\varphi : U^p \to X_\alpha^p$ of X_α^p. Consider now the natural map

$$\nu_1 \times \hat{q}_\varphi : Y^p \times_{X_\varphi^p} S^p \to X_\beta^p \times_{U^p} X_\alpha^p$$

and the isomorphism

$$t_{-s_\beta} \times t_{-s_\alpha} : X_\beta^p \times_{U^p} X_\alpha^p \to X^p \times_{U^p} X^p.$$

Let $\mathscr{P}^p$ denote the pullback

$$\mathscr{P}^p := (\nu_1 \times \hat{q}_\varphi)^* \circ (t_{-s_\beta} \times t_{-s_\alpha})^* (\mathscr{P}_{|X^p \times_{U^p} X^p}),$$

Note that the composite map $(t_{-s_\beta} \times t_{-s_\alpha}) \circ (\nu_1 \times \hat{q}_\varphi) : Y^p \times S^p \to X^p \times_{U^p} X^p$ factors through a small resolution $\widehat{X^p \underset{U^p}{\times} X^p} \to X^p \times_{U^p} X^p$. This implies that the derived pullback of the torsion free sheaf $\mathscr{P}_{|X^p \times_{U^p} X^p}$ via $(t_{-s_\beta} \times t_{-s_\alpha}) \circ (\nu_1 \times \hat{q}_\varphi)$ is equal to the pullback as a coherent sheaf. Indeed, the pullback of $\mathscr{P}_{|X^p \times_{U^p} X^p}$ to the small resolution is locally free and hence it will pullback to a coherent sheaf $\mathscr{P}^p$ (in fact a line bundle) $Y^p \times_{X_\varphi^p} S^p$.

The next step is to observe that without a loss of generality we may assume that the the gerbes ${}_\alpha\mathscr{L}_\beta{}^p \to X_\beta^p$ and ${}_\beta\mathscr{E}_\alpha{}^p \to X_\alpha^p$ are trivial. Explicitly this means that we can find line bundles $\mathscr{Q}^p \in \mathrm{Pic}(Y^p)$ and $\psi^p \in \mathrm{Pic}(S^p)$ on Y^p and S^p respectively, together with isomorphisms

$$\mathscr{P}_{1\text{–}2,\, m\cdot 3|Y^p \times_{X_\varphi^p} Y^p} \cong p_1^* \mathscr{Q}^p \otimes p_2^* (\mathscr{Q}^p)^{-1} \tag{4.3}$$

$$\Phi_{\beta|S^p \times_{X_\alpha^p} S^p} = p_1^*(\psi^p) \otimes p_2^*(\psi^p)^{-1}, \tag{4.4}$$

satisfying the obvious cocycle conditions.

We will work with particular trivializations of ${}_{\alpha}\mathscr{L}_{\beta}{}^{p} \to X^p_\beta$ and ${}_{\beta}\mathscr{E}_{\alpha}{}^{p} \to X^p_\alpha$ which we build in terms of the sections s_φ and s_β respectively. To construct $\mathscr{Q}^p$ we use the rational map

$$Y^p \xrightarrow{\quad\nu\quad} X^p_\varphi \times_{U^p} X^p_\beta \overset{t_{-s_\varphi} \times q^m_\beta}{- - - - - - \dashrightarrow} X^p \times_{U^p} X^p.$$

Note that $(t_{-s_\varphi} \times q^m_\beta) \circ \nu$ is a morphism from the complement of the exceptional curve n^p for the small resolution ν to the complement of the unique singular point in $X^p \times_{U^p} X^p$. Since the standard Poincare sheaf $\mathscr{P}$ is a rank one torsion free sheaf on $X \times_B X$ which fails to be locally free only at the singular points of $X \times_B X$, it follows that the pullback of $\mathscr{P}$ via the map $(t_{-s_\varphi} \times q^m_\beta) \circ \nu$ makes sense as a line bundle defined on $Y^p \setminus n^p$. Combined with the observation that Y^p is smooth and that $n^p \subset Y^p$ is of codimension two, it follows that this pullback extends to a unique line bundle $\mathscr{Q}^p$ on all of Y^p. We can now use the seesaw principle in the same way we did in the proof of Theorem 3.1 to show that the isomorphism (4.3) exists and satisfies the cocycle condition.

Similarly, to construct ψ^p we note that the section s_β gives rise to a relative line bundle $\Sigma_\beta \otimes \mathcal{O}_{X^p_\beta}(-m \cdot s_\beta) \in \mathscr{P}ic^0(X^p_\beta/U_p)$. Under the canonical identification $\mathscr{P}ic^0(X^p_\beta/U^p) = \mathscr{X}^p = \mathscr{P}ic^0(X^p_\varphi/U^p)$, this relative bundle corresponds to a relative line bundle ρ^p of degree zero along the fibers of $X^p_\varphi \to U^p$ and hence to a globally defined line bundle $\widetilde{\psi}^p$ on X^p_φ which restricts to $\rho^p \in \mathscr{P}ic^0(X^p_\varphi/U^p)$. We normalize $\widetilde{\psi}^p$ by choosing a trivialization $s^*_\varphi\widetilde{\psi}^p \cong \mathcal{O}_{U^p}$. Now the argument used in Theorem 3.2 implies that on $(S^p \times_{X^p_\alpha} S^p)_{|U_p \setminus \{p\}}$ we can find an isomorphism $\Phi_\beta \cong p_1^*\widetilde{\psi}^p \otimes p_2^*(\widetilde{\psi}^p)^{-1}$ which (after possibly rescaling the trivialization $s^*_\varphi\widetilde{\psi}^p \cong \mathcal{O}_{U^p}$) will also satisfy the cocycle condition. Now since S is an atlas for the gerbe ${}_{\beta}\mathscr{E}_{\alpha}$ we conclude that $\widetilde{\psi}^p$ on $(S^p \times_{X^p_\alpha} S^p)_{|U_p \setminus \{p\}}$ extends to a unique line bundle ψ^p on $S^p \times_{X^p_\alpha} S^p$ equipped with an isomorphism (4.4) satisfying the cocycle condition.

We now define the coherent sheaf $\Pi^p \in \mathrm{Coh}(Y^p \times_{X^p_\varphi} S^p)$ by setting

$$\Pi^p := \mathscr{P}^p \otimes p_Y^*(\mathscr{Q}^p)^{-1} \otimes p_S^*(\psi^p)^{-1},$$

where $p_Y : Y \times S \to Y$ and $p_S : Y \times S \to S$ denote the natural projections.

With this notation we now have

LEMMA 4.3. *For any $p \in B \setminus B^o$ there is an isomorphism*

$$\Pi^o_{|(Y^o \times_{X^o_\varphi} S^o) \cap (Y^p \times_{X^p_\varphi} S^p)} \cong \Pi^p_{|(Y^o \times_{X^o_\varphi} S^o) \cap (Y^p \times_{X^p_\varphi} S^p)}.$$

In particular the sheaves $\{\Pi^p\}_{p\in B\setminus B^o}$ *glue to the sheaf* Π^o *to yield a globally defined analytic coherent sheaf* Π *on* $Y \times S$.

Proof. We have to show that Π^p is naturally isomorphic to the trivial line bundle on $(Y^o \times_{X^o_\varphi} S^o) \cap (Y^p \times_{X^p_\varphi} S^p)$.

Write U^{po} for the punctured disc $U^p \setminus \{p\}$ and for any space or stack $Z \to B$ write Z^{po} for the fiber product $Z \times_B U^{po}$. Using the isomorphisms t_{-s_φ} and t_{-s_β} we can now identify $\pi_\varphi : X^{po}_\varphi \to U^{po}$ and $\pi_\beta : X^{po}_\beta \to U^{po}$ with the smooth elliptic fibration $\pi : X^{po} \to U^{po}$. Under these identifications the intersection $(Y^o \times_{X^o_\varphi} S^o) \cap (Y^p \times_{X^p_\varphi} S^p)$ gets identified with the fiber product $X^{po} \times_{U^{po}} X^{po}$. Using the same trivializations to recast $\mathscr{P}^p$, $\mathscr{Q}^p$ and ψ^p as sheaves on $X^{po} \times_{U^{po}} X^{po}$ we get that $\mathscr{P}^p$ becomes $p^*_{1,m\cdot 2}\mathscr{P}$, $\mathscr{Q}^p$ becomes $p^*_{m\cdot 1,2}\mathscr{P}$ and ψ^p becomes $\mathcal{O}$. In particular we get that Π^p corresponds to $p_{1,m\cdot 2}\mathscr{P} \otimes p^*_{m\cdot 1,2}\mathscr{P}^{-1}$ and so it suffices to check that

$$p_{1,m\cdot 2}\mathscr{P} \otimes p^*_{m\cdot 1,2}\mathscr{P}^{-1} \cong \mathcal{O}_{X^{po}\times_{U^{po}}X^{po}}.$$

This however follows immediately from the universal property of $\mathscr{P}$ on $X \times_B X$ and the seesaw theorem. $\square$

The sheaf Π gives rise to a well defined integral transform

$$\begin{aligned} D^b(Y) &\xrightarrow{\ \boldsymbol{F}\ } D^b(S) \\ L &\longrightarrow p_{S*}(p_Y^* L \otimes \Pi) \end{aligned} \tag{4.5}$$

between the derived categories of analytic coherent sheaves on Y and S respectively. If in addition $\alpha \in Ш(X) \subset Ш_{an}(X)$ is also algebraic, then the atlases Y and S of the lifting and extension gerbes are proper separated algebraic spaces and so we can invoke the GAGA theorem [**Art70**, Corollary 7.15] to conclude that Π is an algebraic coherent sheaf on $Y \times S$. This implies that in the algebraic context $\boldsymbol{F}$ makes sense as an integral transform between algebraic coherent sheaves.

We now have the following

PROPOSITION 4.4. *The functor* $\boldsymbol{F} : D^b(Y) \to D^b(S)$ *defined by* (4.5) *maps the descent data for the lifting presentation to the descent data for the extension presentation and thus defines a functor* $\boldsymbol{FM} : D^b_1({}_\alpha X_\beta) \to D^b_1({}_\beta X_\alpha)$.

Proof. Let $\mathcal{L}$ be an object in $D^b_1({}_\alpha X_\beta)$ represented by descent datum $(L, \boldsymbol{f})$ for the presentation ${}_\alpha\mathscr{L}_\beta$. In other words, L is an object in $D^b(Y)$ and $\boldsymbol{f} : p_1^* L \to p_2^* L \otimes \mathscr{P}_{1-2,\, m\cdot 3}$ is a quasi-isomorphism on $Y \times_{X_\beta} Y$ satisfying the cocycle condition on $Y \times_{X_\beta} Y \times_{X_\beta} Y$.

Consider the object $\boldsymbol{F}L \in D^b(S)$. To prove the proposition we need to construct a quasi-isomorphism $\boldsymbol{g} : p_1^*\boldsymbol{F}L \to p_2^*\boldsymbol{F}L \otimes \Phi_\beta^{-1}$ on $S \times_{X_\alpha} S$ which depends functorially on $\boldsymbol{f}$ and satisfies the cocycle condition on $S \times_{X_\alpha} S \times_{X_\alpha} S$.

Let $\Gamma := Y \times_{X_\varphi} S$ and let $p_Y : \Gamma \to Y$ and $p_S : \Gamma \to S$ denote the natural projections. Then $\boldsymbol{F}L = p_{S*}(p_Y^*L \otimes \Pi)$, and so our problem boils down to finding a quasi-isomorphism

$$\boldsymbol{g} : p_1^* p_{S*}(p_Y^*L \otimes \Pi) \overset{\cong}{\to} p_2^* p_{S*}(p_Y^*L \otimes \Pi) \otimes \Phi_\beta^{-1}$$

in $D^b(S \times_{X_\alpha} S)$, which depends functorially on $\boldsymbol{f}$. In other words we want to compare the objects $p_1^* p_{S*}(p_Y^*L \otimes \Pi)$ and $p_2^* p_{S*}(p_Y^*L \otimes \Pi)$ on $S \times_{X_\alpha} S$.

Since we have an obvious commutative diagram

(4.6)
$$\begin{array}{ccc}
Y \times_{X_\beta} Y & \overset{p_1}{\underset{p_2}{\rightrightarrows}} & Y \\
{\scriptstyle p_Y \times p_Y}\uparrow & & \uparrow{\scriptstyle p_Y} \\
\Gamma \times_{X_\beta} \Gamma & \overset{\mathrm{p}_1}{\underset{\mathrm{p}_2}{\rightrightarrows}} & \Gamma \\
{\scriptstyle p_S \times p_S}\downarrow & & \downarrow{\scriptstyle p_S} \\
S \times_B S & \overset{\mathrm{pr}_1}{\underset{\mathrm{pr}_2}{\rightrightarrows}} & S \\
{\scriptstyle \iota}\uparrow & & \| \\
S \times_{X_\alpha} S & \overset{p_1}{\underset{p_2}{\rightrightarrows}} & S
\end{array}$$

we see that equivalently we can compare the objects $\iota^* \mathrm{pr}_1^* p_{S*}(p_Y^*L \otimes \Pi)$ and
$\iota^* \mathrm{pr}_2^* p_{S*}(p_Y^*L \otimes \Pi)$. To compute these objects, we would like to perform a base change in the commutative squares

(4.7)
$$\begin{array}{ccc}
\Gamma_{X_\beta}\Gamma & \xrightarrow{\mathrm{p}_1} & \Gamma \\
{\scriptstyle p_S \times p_S}\downarrow & & \downarrow{\scriptstyle p_S} \\
S \times_B S & \xrightarrow[\mathrm{pr}_1]{} & S
\end{array}
\qquad
\begin{array}{ccc}
\Gamma_{X_\beta}\Gamma & \xrightarrow{\mathrm{p}_2} & \Gamma \\
{\scriptstyle p_S \times p_S}\downarrow & & \downarrow{\scriptstyle p_S} \\
S \times_B S & \xrightarrow[\mathrm{pr}_2]{} & S
\end{array} .$$

Unfortunately these squares are not cartesian and so we do not have the base change property on the nose. However we have the following useful

REMARK 4.5. Let X, Y, Z and T be analytic (or algebraic) spaces and let

$$\begin{array}{ccc} Z & \xrightarrow{p} & X \\ {\scriptstyle q}\downarrow & & \downarrow{\scriptstyle f} \\ Y & \xrightarrow[g]{} & T \end{array}$$

be a commutative square of proper maps. Suppose further that the natural map $u : Z \to Y \times_T X$ satisfies $u_* \mathcal{O}_Z = \mathcal{O}_{Y \times_T X}$. Then for every $F \in D^b(X)$ we have a base change identification $q_* p^* F = g^* f_* F$ in $D^b(Y)$. Indeed, we can complete the above square to a commutative diagram

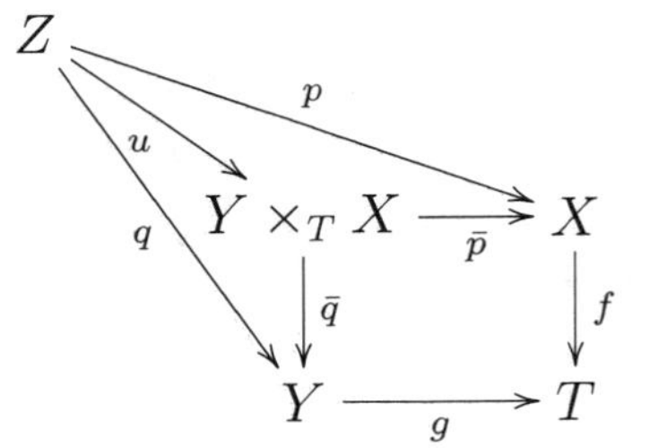

Since $\bar{q}_* \bar{p}^* F = g^* f_* F$ we get

(transitivity)	$q_* p^* F = \bar{q}_* u_* u^* \bar{p}^* F$
(projection formula)	$= \bar{q}_* (\bar{p}^* F \otimes u_* \mathcal{O}_Z)$
(assumption)	$= \bar{q}_* \bar{p}^* F$
(base change)	$= g^* f_* F.$

In view of the previous remark we will be able to treat the squares (4.7) as base change squares if we can show that for the maps

$$\begin{aligned} u_1 &: \Gamma \times_{X_\beta} \Gamma \to (S \times_B S) \times_{p_1, S, ps} \Gamma, \qquad \text{and} \\ u_2 &: \Gamma \times_{X_\beta} \Gamma \to (S \times_B S) \times_{p_2, S, ps} \Gamma \end{aligned}$$

we have $u_{1*} \mathcal{O} = \mathcal{O}$ and $u_{2*} \mathcal{O} = \mathcal{O}$.

To check this, note that

$$\begin{aligned} \Gamma \times_{X_\beta} \Gamma &= (S \times_B S) \times_{X_\varphi \times_B X_\varphi} (Y \times_{X_\beta} Y) \\ (S \times_B S) \times_{p_1, S, ps} \Gamma &= (S \times_B S) \times_{\varepsilon o p_1, X_\varphi, \nu_1} Y \\ (S \times_B S) \times_{p_2, S, ps} \Gamma &= (S \times_B S) \times_{\varepsilon o p_2, X_\varphi, \nu_2} Y, \end{aligned}$$

and so u_i, $i = 1, 2$ are given explicitly by

$$\begin{array}{rccc} u_i : & (S \times_B S) \times_{X_\varphi \times_B X_\varphi} (Y \times_{X_\beta} Y) & \longrightarrow & (S \times_B S) \times_{\varepsilon o p_i, X_\varphi, \nu_1} Y \\ & ((a_1, a_2), (b_1, b_2)) & \longrightarrow & ((a_1, a_2), b_i). \end{array}$$

This implies in particular that the maps u_i fit in the cartesian squares

$$\begin{array}{ccc} (S\times_B S) \underset{X_\varphi\times_B X_\varphi}{\times} (Y\times_{X_\beta} Y) & \xrightarrow{\quad p_2\circ p_{Y\times X_\beta}\quad} & Y \\ \downarrow{\scriptstyle u_1} & & \downarrow{\scriptstyle \nu} \\ (S\times_B S)\times_{\varepsilon\circ p_1, X_\varphi,\nu_1} Y & \xrightarrow[(\varepsilon\circ p_2\circ \mathrm{pr}_{S\times_B S})\times(\nu_2\circ\mathrm{pr}_Y)]{} & X_\varphi\times_B X_\beta \end{array}$$

and

$$\begin{array}{ccc} (S\times_B S) \underset{X_\varphi\times_B X_\varphi}{\times} (Y\times_{X_\beta} Y) & \xrightarrow{\quad p_1\circ p_{Y\times X_\beta}\quad} & Y \\ \downarrow{\scriptstyle u_2} & & \downarrow{\scriptstyle \nu} \\ (S\times_B S)\times_{\varepsilon\circ p_2, X_\varphi,\nu_1} Y & \xrightarrow[(\varepsilon\circ p_1\circ \mathrm{pr}_{S\times_B S})\times(\nu_2\circ\mathrm{pr}_Y)]{} & X_\varphi\times_B X_\beta \end{array}$$

where $\nu : Y \to X_\varphi \times_B X_\beta$ is the small resolution map defining Y.

Since the pullback of $\mathcal{O}$ by any morphism is again $\mathcal{O}$, it suffices to check that there is a canonical isomorphism $\nu_*\mathcal{O}_Y = \mathcal{O}_{X_\varphi\times_B X_\beta}$. This is obvious by the cohomology and base change theorem. Thus $u_{i*}\mathcal{O} = \mathcal{O}$ and so the squares (4.7) have the base change property.

In particular, in $D^b(S\times_B S)$ we get identifications:

$$\mathrm{pr}_i^*\, p_{S*}(p_Y^* L\otimes \Pi) = (p_S\times p_S)_*\, \mathrm{p}_i^*(p_Y^*L\otimes\Pi),$$

for all $L\in D^b(Y)$ and for $i = 1, 2$. Furthermore since

$$(S\times_{X_\alpha} S) \underset{S\times_B S}{\times} (\Gamma\times_{X_\beta}\Gamma) = \Gamma \underset{X_\alpha\times_B X_\beta}{\times} \Gamma,$$

we can identify $\iota^*(p_S\times p_S)_*\,\mathrm{p}_i^*(p_Y^*L\otimes\Pi)$ with $p_{S\times_{X_\alpha}S*}p_i^*(p_Y^*L\otimes\Pi)$ where $p_{S\times_{X_\alpha}S} : \Gamma\times_{X_\alpha\times_B X_\beta}\Gamma \to S\times_{X_\alpha} S$ is the natural projection.

Now, using the commutativity of the top double square in (4.6) we get

$$\begin{aligned} (p_{S\times_{X_\alpha}S})_*p_1^*(p_Y^*L\otimes\Pi) &= (p_{S\times_{X_\alpha}S})_*(((p_Y\times p_Y)^*p_1^*L)\otimes(p_1^*\Pi)), \text{ and}\\ (p_{S\times_{X_\alpha}S})_*p_2^*(p_Y^*L\otimes\Pi) &= (p_{S\times_{X_\alpha}S})_*(((p_Y\times p_Y)^*p_2^*L)\otimes(p_2^*\Pi)). \end{aligned}$$

If in addition $L\in D^b(Y)$ is part of a descend datum $(L, \boldsymbol{f})$ defining an object in $D^b_1({}_\alpha X_\beta)$, we can use $\boldsymbol{f}$ to obtain an identification

$$\begin{aligned} &(p_{S\times_{X_\alpha}S})_*p_1^*(p_Y^*L\otimes\Pi)\\ &\quad = (p_{S\times_{X_\alpha}S})_*((p_Y\times p_Y)^*p_2^*L\otimes((p_Y\times p_Y)^*\mathscr{P}_{1-2,\,m\cdot 3})\otimes(p_1^*\Pi)). \end{aligned}$$

Also

$$\begin{aligned} &((p_{S\times_{X_\alpha}S})_*p_2^*(p_Y^*L\otimes\Pi))\otimes\Phi_\beta^{-1}\\ &\quad = (p_{S\times_{X_\alpha}S})_*(((p_Y\times p_Y)^*p_2^*L)\otimes((p_{S\times_{X_\alpha}S}^*\Phi_\beta^{-1})\otimes(p_2^*\Pi)), \end{aligned}$$

and so, in order to get the desired isomorphism $\boldsymbol{g} : p_1^*\boldsymbol{F}L \to p_2^*\boldsymbol{F}L \otimes \Phi_\beta^{-1}$ in $D^b(S \times_{X_\alpha} S)$ we only have to construct a canonical identification

$$p_1^*\Pi \otimes (p_Y \times p_Y)^* \mathscr{P}_{1\text{–}2,\, m\cdot 3} = p_2^*\Pi \otimes p_{S\times_{X_\alpha}S}^* \Phi_\beta^{-1} \tag{4.8}$$

of coherent sheaves on $\Gamma \times_{X_\alpha \times_B X_\beta} \Gamma$.

We will construct the desired map (4.8) by gluing some locally defined but canonical identifications. Note that at this point we have completely eliminated the derived category from the picture. In particular, we are left with a question about sheaves, not complexes, and so gluing is a relatively simple matter.

(i) Over the part of $\Gamma \times_{X_\alpha \times_B X_\beta} \Gamma$ sitting over $B^o \subset B$, the identification (4.8) is just the one established in section 4. Indeed, by definition $\Pi_{|\Gamma^o} = \mathscr{O}_{\Gamma^o}$ and so constructing (4.8) over B^o becomes equivalent to constructing a canonical identification (3.13). Such a construction was carried out in the proof of Proposition 3.4.

(ii) Over the part of $\Gamma \times_{X_\alpha \times_B X_\beta} \Gamma$ sitting over $U^p \subset B$, $p \in B \setminus B^o$, we can use again the line bundles $\mathscr{Q}^p \to Y^p$ and $\psi^p \to S^p$ appearing in the construction of $\Pi^p = \Pi_{|\Gamma^p}$ to trivialize our gerbes. Recall that $\mathscr{Q}^p$ and ψ^p come equipped with the natural isomorphisms (4.3) and (4.4). Furthermore $\Pi^p = \mathscr{P}^p \otimes p_Y^*(\mathscr{Q}^p)^{-1} \otimes p_S^*(\psi^p)^{-1}$ and so establishing (4.8) over $\Gamma^p \times_{X_\alpha^p \times_{U^p} X_\beta^p} \Gamma^p$ reduces to constructing an identification:

$$\begin{array}{c} p_1^*\mathscr{P}^p \otimes p_1^*p_Y^*(\mathscr{Q}^p)^{-1} \otimes p_1^*p_S^*(\psi^p)^{-1} \otimes p_1^*p_Y^*\mathscr{Q}^p \otimes p_2^*p_Y^*(\mathscr{Q}^p)^{-1} \\ \| \\ p_2^*\mathscr{P}^p \otimes p_2^*p_Y^*(\mathscr{Q}^p)^{-1} \otimes p_2^*p_S^*(\psi^p)^{-1} \otimes p_1^*p_S^*(\psi^p)^{-1} \otimes p_2^*p_S^*\psi^p \end{array}$$

or equivalently, after the obvious cancellations, an identification

$$p_1^*\mathscr{P}^p = p_2^*\mathscr{P}^p.$$

However $\mathscr{P}^p \in \operatorname{Coh}(\Gamma^p)$ was defined as a pullback of a sheaf on $X_\alpha^p \times_{U^p} X_\beta^p$ and so we get a canonical identification $p_1^*\mathscr{P}^p = p_2^*\mathscr{P}^p$ on $\Gamma^p \times_{X_\alpha^p \times_{U^p} X_\beta^p} \Gamma^p$. This yields the desired canonical identification (4.8) over the part of $\Gamma \times_{X_\alpha \times_B X_\beta} \Gamma$ sitting over U^p

Finally it only remains to observe that the isomorphisms chosen in (4.3) and (4.4) were the ones used in the proof of Lemma 4.3 to glue Π^p and Π^o on the overlap $\Gamma^p \cap \Gamma^o$. Therefore the identifications in items **(i)** and **(ii)** above glue on the overlaps $(\Gamma \times_{X_\alpha \times_B X_\beta} \Gamma) \times_B U^{po}$ and so we have found our global identification (4.8). This finishes the proof of the proposition. □

We are now ready to complete the

Proof of Theorem A. The only thing left to show is that the gerby Fourier-Mukai transform $\boldsymbol{FM} : D^b_1({}_\alpha X_\beta) \to D^b_1({}_\beta X_\alpha)$ constructed in Proposition 4.4 is an equivalence of categories. We will again use the criterion of Bondal-Orlov and Bridgeland applied to the spanning class Ω of gerby points in ${}_\alpha X_\beta$ described in Claim 3.7. As before we need to show that $\boldsymbol{FM}$ intertwines the Serre functors on the sheaves $\mathcal{O}_x \in \Omega$ and that $\boldsymbol{FM}$ satisfies the orthogonality property

$$\boldsymbol{FM} : \mathrm{Hom}^i_{D^b_1({}_\alpha X_\beta)}(\mathcal{O}_{x_1}, \mathcal{O}_{x_2}) \widetilde{\to} \mathrm{Hom}^i_{D^b_1({}_\beta X_\alpha)}(\boldsymbol{F}\mathcal{O}_{x_1}, \boldsymbol{F}\mathcal{O}_{x_2}),$$

for all $i \in \mathbb{Z}, x_1, x_2 \in X_\beta$.

Since by definition the Fourier-Mukai image $\boldsymbol{FM}\mathcal{O}_x$ is supported on the fiber $({}_\beta X_\alpha)_b$ of ${}_\beta X_\alpha$ over the point $b = \pi_\beta(x) \in B$, it suffices to check the intertwining and orthogonality properties of $\boldsymbol{FM}$ locally in the base B.

Over B^o these properties were established in Claims 3.8 and 3.9. To check the properties for the parts of our gerbes sitting over $U^p \subset B$ we note that the proof of Lemma 4.3 shows that over U^p the functor $\boldsymbol{FM}$ fits in a commutative diagram of functors

$$\begin{array}{ccc}
D^b_1({}_\alpha X_\beta{}^p) & \xrightarrow{\ \boldsymbol{FM}\ } & D^b_{-1}({}_\beta X_\alpha{}^p) \\
{\scriptstyle \nu_1^*(\bullet)\otimes \mathscr{Q}^p}\uparrow & & \uparrow{\scriptstyle \nu_2^*(\bullet)\otimes(\psi^p)^{-1}} \\
D^b(X^p_\beta) & & D^b(X^p_\alpha) \\
{\scriptstyle t^*_{-s_\beta}}\uparrow & & \uparrow{\scriptstyle t^*_{-s_\alpha}} \\
D^b(X^p) & \xrightarrow[p_{2*}(p_1^*(\bullet)\otimes\mathscr{P})]{} & D^b(X^p)
\end{array}$$

where the vertical arrows are equivalences. However the bottom arrow is the usual integral transform with respect to the Poincare sheaf on an elliptic surface having at most I_1 fibers. Such a transform is an equivalence, e.g. by [**BM02**]. Finally, the functor $(\nu_1^*(\bullet) \otimes \mathscr{Q}^p) \circ t^*_{-s_\beta}$ transforms a structure sheaf of a point $x \in X^p$ into a sheaf in the spanning class Ω^p for ${}_\alpha X_\beta{}^p$ and clearly every sheaf in Ω^p is obtained this way. This implies that $\boldsymbol{FM}$ has the orthogonality and intertwining properties for sheaves in Ω^p. The theorem is proven, □

From the statement of Theorem A one can derive a whole sequence of new cases of Căldăraru's conjecture. Indeed, suppose X is an elliptic K3 surface whose singular fibers are of type I_1 only. Note that for any element $\alpha \in \text{Ш}(X) = \text{Hom}(\boldsymbol{T}_X, \mathbb{Q}/\mathbb{Z})$ we have a natural Hodge isometry $\boldsymbol{T}_{X_\alpha} \cong \ker(\alpha)$ induced by the isogeny of X_α and X. In terms of the identifications $\text{Ш}(X) = Br'(X) = \text{Hom}(\boldsymbol{T}_X, \mathbb{Q}/\mathbb{Z})$ and $Br'(X_\alpha) = \text{Hom}(\ker(\alpha), \mathbb{Q}/\mathbb{Z})$, the surjective map $T_{alpha} : \text{Ш}(X) \to Br'(X_\alpha)$ sends a homomorphism $a : \boldsymbol{T}_X \to \mathbb{Q}/\mathbb{Z}$ to its restriction $a_{|\ker(\alpha)} : \ker(\alpha) \to \mathbb{Q}/\mathbb{Z}$. Now we have:

Corollary 4.6. *Let X be an elliptic K3 surface whose singular fibers are of type I_1 only. Let $\alpha, a \in \text{Ш}(X) = \text{Hom}(\boldsymbol{T}_X, \mathbb{Q}/\mathbb{Z})$ and let $(b, \beta) \in \text{Ш}(X)^{\times 2}$ be in the $SL(2,\mathbb{Z})$ orbit of (α, a). Then*

(a) $\ker \left[\ker(\alpha) \xrightarrow{a} \mathbb{Q}/\mathbb{Z}\right]$ *and* $\ker \left[\ker(\beta) \xrightarrow{b} \mathbb{Q}/\mathbb{Z}\right]$ *are Hodge isometric;*

(b) $D^b_1({}_aX_\alpha)$ *and* $D^b_1({}_bX_\beta)$ *are equivalent.*

Proof. Part (b) follows from the fact that for every (a, α) we have equivalences $D^b_1({}_aX_\alpha) \cong D^b_1({}_{-\alpha}X_a)$ (by Theorem A) and $D^b_1({}_aX_\alpha) \cong D^b_1({}_{a+\alpha}X_\alpha)$ (since $T_\alpha(\alpha) = 0$).

For part (a) observe that by our identifications we have isometries of Hodge lattices

$$\ker \left[\ker(\alpha) \xrightarrow{a} \mathbb{Q}/\mathbb{Z}\right] = \ker(a) \cap \ker(\alpha), \text{ and}$$

$$\ker \left[\ker(\beta) \xrightarrow{b} \mathbb{Q}/\mathbb{Z}\right] = \ker(b) \cap \ker(\beta).$$

Since $(a, \alpha) : \boldsymbol{T}_X \to (\mathbb{Q}/\mathbb{Z})^2$ and $(b, \beta) : \boldsymbol{T}_X \to (\mathbb{Q}/\mathbb{Z})^2$ differ by an element of $\text{SL}(2,\mathbb{Z})$ acting on $(\mathbb{Q}/\mathbb{Z})^2$, it follows that $\ker(a) \cap \ker(\alpha) = \ker(b) \cap \ker(\beta)$ as sublattices in $\boldsymbol{T}_X$. □

CHAPTER 5

Modified T-duality and the SYZ conjecture

The celebrated work of Strominger, Yau and Zaslow [**SYZ96**] interprets mirror symmetry of Calabi-Yaus in terms of special Lagrangian (SLAG) torus fibrations. If a CY manifold X (with "large complex struture") has mirror X', [**SYZ96**] conjecture the existence of fibrations $\pi : X \to B$ and $\pi' : X' \to B$ whose generic fibers are SLAG tori dual to each other: each parameterizes $U(1)$ flat connections on the other. In particular, each of these fibrations admits a SLAG zero-section, corresponding to the trivial connection on the dual fibers. The analogy with the situation considered in the main part of our work is clear: the SLAG torus fibrations replace the elliptic fibrations, and mirror symmetry (interchanging D-branes of type B with D-branes of type A) replaces the Fourier-Mukai transform (which interchanges vector bundles with spectral data).

In this context, the analogue of our gerbes and the Brauer group is given by the "B-fields" $\alpha \in H^2(X, \mathbb{R}/\mathbb{Z})$. On the other hand, the SLAG analogue of the Tate-Shafarevich group of X' is given by $H^1(B, X')$, which over the locus where π, π' are smooth can be identified with $H^1(B, R^1\pi_*(\mathbb{R}/\mathbb{Z}))$. As in the proof of lemma 2.11, $H^2(X, \mathbb{R}/\mathbb{Z})$ is related via a Leray spectral sequence to the three groups $H^i(B, R^{2-i}\pi_*(\mathbb{R}/\mathbb{Z}))$, for $i = 1, 2, 3$. Now for $i = 0$, the local system $H^0(B, R^2\pi_*(\mathbb{R}/\mathbb{Z}))$ can be identified, over the locus where π is smooth, with the group of homotopy classes of sections of $X \to B$. Therefore, if the fibration $X \to B$ is good in the sense of [**Gro98, Gro99**], $H^0(B, R^2\pi_*(\mathbb{R}/\mathbb{Z}))$ should be thought of as the analogue of the Mordell-Weil group of $X \to B$. Assume that the SLAG fibration $X \to B$ is generic, in the sense that the local system $R^2\pi_*(\mathbb{R}/\mathbb{Z})$ has no global sections.

The Leray spectral sequence therefore gives a Brauer-to-Tate-Shafarevich map:

$$H^2(X, \mathbb{R}/\mathbb{Z}) \to H^1(B, R^1\pi_*(\mathbb{R}/\mathbb{Z})).$$

We therefore may as well start with a pair of B-fields $\alpha \in H^2(X, \mathbb{R}/\mathbb{Z}), \beta \in H^2(X', \mathbb{R}/\mathbb{Z})$ on X and X' respectively. Since mirror symmetry involves the B-field in an essential way, this suggests that the SYZ conjecture, which is a SLAG analogue of the Fourier-Mukai transform for elliptic fibrations, must be supplemented by an analogue of our Theorem A. Let $\mathcal{M}$ be the CY moduli space on which mirror symmetry acts. The emerging picture is that $\mathcal{M}$ looks, at least in some approximate sense, like an integrable system. The base is a real submanifold $\mathcal{M}_{\mathbb{R}} \subset \mathcal{M}$. The normal directions are parametrized by the B-fields α, β , so they form a torus isomorphic to $H^2(X, \mathbb{R}/\mathbb{Z}) \times H^2(X', \mathbb{R}/\mathbf{Z})$. The original SYZ conjecture holds along $\mathcal{M}_{\mathbb{R}}$. As we move in normal directions, X becomes gerby along B-directions on X, while along B-directions on X', the SLAG fibration on X loses its SLAG zero-section. Mirror symmetry interchanges these two behaviors.

Indeed, the moduli space $\mathcal{M}$ parameterizes pairs (X, α) where X is a complex manifold together with a Calabi-Yau metric, and $\alpha \in H^2(X, \mathbb{R}/\mathbb{Z})$ is a B-field on it. Mirror symmetry is an involution $\boldsymbol{MS} : \mathcal{M} \to \mathcal{M}$, presumably defined in a neighborhood of the "large complex structure" point, taking (X, α) to (X', β). We let $\mathcal{M}_{\mathbb{R}} \subset \mathcal{M}$ denote the locus where $\alpha = \beta = 0$. It is a component of the fixed locus of the antilinear involution which reverses the signs of α and β. We denote a point of $\mathcal{M}_{\mathbb{R}}$ by the mirror pair X, X'. Now a B-field $\beta \in H^2(X', \mathbb{R}/\mathbb{Z})$ on X' determines a point (X', β) of $\mathcal{M}$, hence a mirror point $X_\beta := \boldsymbol{MS}(X', \beta)$. For small β, this X_β is a deformation of X, so the additional B-field $\alpha \in H^2(X, \mathbb{R}/\mathbb{Z})$ on X determines a corresponding B-field $T_\beta(\alpha) \in H^2(X_\beta, \mathbb{R}/\mathbb{Z})$ on X_β.

CONJECTURE 5.1.

- The SYZ picture holds (near the large complex structure/large volume limit) on $\mathcal{M}_{\mathbb{R}}$.
- For a B-field B' on X', the deformed Calabi-Yau $X_{B'}$ admits a SLAG fibration (generally *without* a section) whose Jacobian (i.e. double dual) is the original SLAG fibration (with section) on X.
- Mirror symmetry preserves the integrable system structure: for any pair α, β, the mirror of $(X_\beta, T_\beta(\alpha))$ is $(X'_\alpha, T_\alpha(\beta))$.

We note that this modification of [**SYZ96**] is consistent with recent interpretations in the literature (see [**Hit01**] and references therein) of D-branes on Calabi-Yaus in the presence of a B-field: A D-brane of type B on X is a coherent sheaf on the gerbe given by the B-filed β, while a D-brane of type A on X' is, roughly, a flat $U(1)$ connection on the restriction of the gerbe α to a SLAG submanifold in X'. The third part of this conjecture is the exact SLAG translation of Theorem A.

APPENDIX A

Duality for representations of 1-motives, by Dmitry Arinkin

In this appendix, we sketch a different approach to the Fourier-Mukai transform for $\mathcal{O}^\times$-gerbes over smooth genus one fibrations (Theorem B). In this approach, Theorem B claims that the dual commutative group stacks (of a certain type) have equivalent derived categories of coherent sheaves. Let us review the duality for commutative group stacks (sometimes called the duality for generalized 1-motives).

Recall that the dual $X^\vee$ of an abelian variety X is the moduli space of line bundles with zero first Chern class on X. Equivalently, $X^\vee$ parametrizes the extensions of the algebraic group X by $\mathbb{G}_m$. In this form, the definition immediately generalizes to stacks: for a commutative group stack $\mathscr{X}$, its dual $\mathscr{X}^\vee$ is the moduli stack of extensions of commutative group stacks

$$1 \to \mathbb{G}_m \to G \to \mathscr{X} \to 0.$$

The sum of extensions defines a group operation on $\mathscr{X}^\vee$; actually, $\mathscr{X}^\vee$ is naturally a commutative group stack.

REMARK A.1. For technical reasons, we use a slightly different definition of the dual stack (Definition A.1). This allows to avoid the discussion of short exact sequences of group stacks; also, the group structure on $\mathscr{X}^\vee$ seems somewhat more natural.

Let $\mathscr{P} \to \mathscr{X}^\vee \times \mathscr{X}$ be the universal $\mathscr{X}^\vee$-family of extensions of $\mathscr{X}$ by $\mathbb{G}_m$; in particular, $\mathscr{P}$ is a $\mathbb{G}_m$-torsor on $\mathscr{X}^\vee \times \mathscr{X}$ (in fact, $\mathscr{P}$ is a biextension of $\mathscr{X}^\vee \times \mathscr{X}$ by $\mathbb{G}_m$). Notice that we can also view $\mathscr{P}$ as a $\mathscr{X}$-family of extensions of $\mathscr{X}^\vee$ by $\mathbb{G}_m$; this defines a morphism $\mathscr{X} \to (\mathscr{X}^\vee)^\vee$. The main idea of the Fourier-Mukai transform for commutative group stacks can be informally stated as follows:

(A.1) For a "good" commutative group stack $\mathscr{X}$, the morphism $\mathscr{X} \to (\mathscr{X}^\vee)^\vee$ is an isomorphism, and the Fourier-Mukai transform defined by $\mathscr{P}_\mathbb{C}$ is an equivalence $\boldsymbol{FM} : D^b(\mathscr{X}) \to D^b(\mathscr{X}^\vee)$. Here $\mathscr{P}_\mathbb{C}$ is the line bundle on $\mathscr{X}^\vee \times \mathscr{X}$ associated to the $\mathbb{G}_m$-torsor $\mathscr{P}$.

Now let us explain how Theorem B fits into the framework of the duality for commutative group stacks. First, we notice that the $\mathcal{O}^\times$-gerbe ${}_\alpha X_0$ over X is a group stack. Then we see that ${}_\alpha X_\beta$ is a torsor over the group stack ${}_\alpha X_0$; more precisely, the gerbes constructed using the lifting presentation and the extension presentation (from Section 3.1) have a natural structure of ${}_\alpha X_0$-torsors.

Torsors over a group stack can be thought of as extensions of $\mathbb{Z}$ by this group stack; in this way, ${}_\alpha X_\beta$ defines a commutative group stack ${}_\alpha \widetilde{X}_\beta$ that fits into an exact sequence

$$0 \to {}_\alpha X_0 \to {}_\alpha \widetilde{X}_\beta \to \mathbb{Z} \to 0.$$

The argument in section 4 shows that the constructions of the lifting presentation and the extension presentation are dual, so ${}_\alpha \widetilde{X}_\beta$ and ${}_{-\beta}\widetilde{X}_\alpha$ are dual commutative group stacks (provided that we use the lifting presentation for one of the two stacks and the extension presentation for the other). Moreover, these stacks are "good" in the sense of (A.1), and so the Fourier-Mukai transform gives an equivalence between $D^b({}_\alpha \widetilde{X}_\beta)$ and $D^b({}_{-\beta}\widetilde{X}_\alpha)$. The Fourier-Mukai transform of Theorem B is the restriction of this equivalence to direct summands in the derived categories (see Section 2)

In the rest of the appendix, we discuss the notion of the dual of a group stack (Section 1) and the special case when the group stack is an extension of $\mathbb{Z}$ (Section 2). No proofs are given, but most statements are almost obvious.

I learned about the duality for commutative group stacks from A. Beilinson, and I am deeply grateful to him for the explanation.

1. Duality for commutative group stacks

From now on, the word 'stack' means an algebraic stack locally of finite type over a fixed base scheme B. All results also have an analytic version.

DEFINITION A.1. For a commutative group stack $\mathscr{X}$, the dual stack $\mathscr{X}^\vee$ parametrizes 1-morphisms of commutative group stacks from $\mathscr{X}$ to $B\mathbb{G}_m$ (the classifying stack of $\mathbb{G}_m$). Thus, for a B-scheme S, the category $\mathscr{X}^\vee(S)$ is the category of 1-morphisms of commutative group S-stacks $\mathscr{X} \times_B S \to B\mathbb{G}_m \times S$. Notice that $\mathscr{X}^\vee$ does not have to be algebraic.

REMARK A.2. For the definition to make sense, we need certain smallness assumptions. Indeed, if $\mathscr{X}$ and $\mathscr{Y}$ are stacks on a site B, the 1-morphisms from $\mathscr{X}$ to $\mathscr{Y}$ form a stack only if $\mathscr{X}$, $\mathscr{Y}$, and B are small. However, this problem can be avoided if we assume that $\mathscr{X}$ is an algebraic stack which is locally of finite type and replace the category of finitely presented B-schemes by an equivalent small category.

EXAMPLE A.1. If $\mathscr{X}$ is an abelian scheme over B, then $\mathscr{X}^\vee$ is the dual abelian scheme.

EXAMPLE A.2. Let $\mathscr{X} = G$ be an affine (or ind-affine) abelian group (over $\mathbb{C}$). Then $\mathscr{X}^\vee$ is the classifying stack of the Cartier dual of G. In particular, if $\mathscr{X} = \mathbb{Z}$, we have $\mathscr{X}^\vee = B\mathbb{G}_m$.

Another example is provided by the stacks ${}_\alpha\widetilde{X}_\beta$ (constructed using either the lifting presentation or the extension presentation). It is clear from the construction that locally on B, the stack ${}_\alpha\widetilde{X}_\beta$ is isomorphic to $X \times B\mathbb{G}_m \times \mathbb{Z}$; globally, it carries a natural filtration $0 \subset \widetilde{X}^{(1)} \subset \widetilde{X}^{(2)} \subset {}_\alpha\widetilde{X}_\beta$ with $\widetilde{X}^{(1)} = B\mathbb{G}_m$, $\widetilde{X}^{(2)}/\widetilde{X}^{(1)} = X$, and ${}_\alpha\widetilde{X}_\beta/\widetilde{X}^{(2)} = \mathbb{Z}$. This implies the following statement:

PROPOSITION A.1. *${}_\alpha\widetilde{X}_\beta$ is "good" in the sense of (A.1).*

Proof. The property of being "good" is local on B, so it is enough to notice that the stacks X, $B\mathbb{G}_m$, and $\mathbb{Z}$ are "good". □

In particular, we see that the Fourier-Mukai transform gives an equivalence between $D^b({}_\alpha\widetilde{X}_\beta)$ and $D^b({}_{-\beta}\widetilde{X}_\alpha)$.

2. Duality for torsors

Now suppose $\mathscr{X}$ is a commutative group stack which is "good", and let $\mathscr{X}'$ be a torsor over $\mathscr{X}$. Denote by $\widetilde{\mathscr{X}}$ the corresponding extension of $\mathbb{Z}$ by $\mathscr{X}$: it fits into the exact sequence

$$0 \to \mathscr{X} \to \widetilde{\mathscr{X}} \to \mathbb{Z} \to 0$$

and $\mathscr{X}'$ is identified with the preimage of $1 \in \mathbb{Z}$. Notice that locally on B, the torsor is trivial, so $\widetilde{\mathscr{X}}$ is isomorphic to $\mathscr{X} \times \mathbb{Z}$. Since both $\mathscr{X}$ and $\mathbb{Z}$ are "good", so is $\widetilde{\mathscr{X}}$. The dual stack $\widetilde{\mathscr{X}}^\vee$ is isomorphic to $\mathscr{X}^\vee \times B\mathbb{G}_m$ locally on B (globally, it contains a substack isomorphic to $B\mathbb{G}_m$, and the quotient equals $\mathscr{X}^\vee$). In particular, if $\mathscr{X}^\vee$ is actually a space (rather than a stack), then $\widetilde{\mathscr{X}}^\vee$ is an $\mathcal{O}^\times$-gerbe over the space. The following statement is clear:

PROPOSITION A.2. *The Fourier-Mukai functor*

$$\boldsymbol{FM} : D^b(\widetilde{\mathscr{X}}) \to D^b(\widetilde{\mathscr{X}}^\vee)$$

induces an equivalence $D^b(\mathscr{X}') \to D^b_1(\widetilde{\mathscr{X}}^\vee)$, where $D^b_1(\widetilde{\mathscr{X}}^\vee)$ is the full subcategory in $D^b(\widetilde{\mathscr{X}}^\vee)$ of objects F such that the action of $\mathbb{G}_m$ on

$H^i(F)$ is tautological. Here the action is induced by the morphism $B\mathbb{G}_m \to \widetilde{\mathscr{X}}^\vee$.

In the case of duality between ${}_\alpha\widetilde{X}_\beta$ and ${}_{-\beta}\widetilde{X}_\alpha$, both stacks are torsors (over ${}_\alpha X_\beta$ and ${}_{-\beta}X_\alpha$, respectively), and so we get equivalences

$$D^b({}_\alpha X_\beta) \to D^b_1({}_{-\beta}\widetilde{X}_\alpha)$$

and

$$D^b_1({}_\alpha\widetilde{X}_\beta) \to D^b({}_{-\beta}X_\alpha).$$

They induce an equivalence

$$D^b({}_\alpha X_\beta) \cap D^b_1({}_\alpha\widetilde{X}_\beta) = D^b_1({}_\alpha X_\beta) \to D^b_1({}_{-\beta}X_\alpha).$$

This is exactly what Theorem B claims.

Bibliography

[AD98] P. Aspinwall and R. Donagi. The heterotic string, the tangent bundle and derived categories. *Adv. Theor. Math. Phys.*, 2(5):1041–1074, 1998.

[Art70] M. Artin. Algebraization of formal moduli II. Existence of modifications. *Ann. of Math.*, 91:88–135, 1970.

[Art74] M. Artin. Versal deformations and algebraic stacks. *Invent. math.*, 27:165–189, 1974.

[BBRP98] C. Bartocci, U. Bruzzo, D. Hernández Ruipérez, and J. Muñoz Porras. Mirror symmetry on $K3$ surfaces via Fourier-Mukai transform. *Comm. Math. Phys.*, 195(1):79–93, 1998.

[BJPS97] M. Bershadsky, A. Johansen, T. Pantev, and V. Sadov. On four-dimensional compactifications of F-theory. *Nuclear Phys. B*, 505(1-2):165–201, 1997.

[BK90] A. Bondal and M. Kapranov. Representable functors, Serre functors and mutations. *Math. USSR. Izv.*, 35:519–541, 1990.

[BKR01] T. Bridgeland, A. King, and M. Ried. Mukai implies McKay: the McKay correspondence as an equivalence of derived categories. *J. Amer. Math. Soc.*, 14:535–554, 2001. math.AG/9908027.

[BM02] T. Bridgeland and A. Maciocia. Fourier-Mukai transforms for $K3$ and elliptic fibrations. *J. Algebraic Geom.*, 11(4):629–657, 2002.

[BM03] V. Brinzanescu and R. Moraru. Holomorphic rank-2 vector bundles on non-Kahler elliptic surfaces, 2003, arXiv:math.AG/0306191.

[BM05] V. Brinzanescu and R. Moraru. Stable bundles on non-Kahler elliptic surfaces, Comm. Math. Phys. **254** (2005), no. 3, 565–580.

[BO95] A. Bondal and D. Orlov. Semiorthogonal decomposition for algebraic varieties, 1995, arXiv:alg-geom/9506012.

[Bre90] L. Breen. Bitorseurs et cohomologie non abélienne. In *The Grothendieck Festschrift, Vol. I*, pages 401–476. Birkhäuser Boston, Boston, MA, 1990.

[Bre94] L. Breen. On the classification of 2-gerbes and 2-stacks. *Astérisque*, 225:160, 1994.

[Bri98] T. Bridgeland. Fourier-Mukai transforms for elliptic surfaces. *J. Reine Angew. Math.*, 498:115–133, 1998.

[Bri99] T. Bridgeland. Equivalences of triangulated categories and Fourier-Mukai transforms. *Bull. London Math. Soc.*, 31:25–34, 1999. math.AG/9809114.

[Bri02] T. Bridgeland. Flops and derived categories. *Invent. Math.*, 147(3):613–632, 2002.

[Bry93] J.-L. Brylinski. *Loop spaces, characteristic classes and geometric quantization*, volume 107 of *Progress in Mathematics*. Birkhäuser Boston Inc., Boston, MA, 1993.

[Căl00] A. Căldăraru. *Derived categories of twisted sheaves on Calabi-Yau manifolds*. PhD thesis, Cornell University, 2000.

[Căl02a] A. Căldăraru. Non-Fine Moduli Spaces of Sheaves on $K3$ Surfaces, Int. Math. Res. Not. 2002, no. 20, 1027–1056.

[Căl02b] A. Căldăraru. Derived categories of twisted sheaves on elliptic threefolds. *J. Reine Angew. Math.*, 544:161–179, 2002.

[Del74] P. Deligne. Théorie de Hodge. III. *Inst. Hautes Études Sci. Publ. Math.*, 44:5–77, 1974.

[Del90] P. Deligne. Catégories tannakiennes. In *The Grothendieck Festschrift, Vol. II*, pages 111–195. Birkhäuser Boston, Boston, MA, 1990.

[DG94] I. Dolgachev and M. Gross. Elliptic threefolds. I. Ogg-Shafarevich theory. *J. Algebraic Geom.*, 3(1):39–80, 1994.

[DG02] R. Donagi and D. Gaitsgory. The gerbe of Higgs bundles. *Transform. Groups*, 7(2):109–153, 2002, arXiv:math.AG/0005132.

[Don97] R. Donagi. Principal bundles on elliptic fibrations. *Asian J. Math.*, 1(2):214–223, 1997.

[Don99] R. Donagi. Heterotic F-theory duality. In *XIIth International Congress of Mathematical Physics (ICMP '97) (Brisbane)*, pages 206–213. Internat. Press, Cambridge, MA, 1999.

[DOPW01a] R. Donagi, B. Ovrut, T. Pantev, and D. Waldram. Standard-model bundles. *Adv. Theor. Math. Phys.*, 5(3):563–615, 2001.

[DOPW01b] R. Donagi, B. Ovrut, T. Pantev, and D. Waldram. Standard models from heterotic M-theory. *Adv. Theor. Math. Phys.*, 5(1):93–137, 2001.

[EHKV01] D. Edidin, B. Hassett, A. Kresch, and A. Vistoli. Brauer groups and quotient stacks. *Amer. J. Math.*, 123(4):761–777, 2001.

[FMW97] R. Friedman, J. Morgan, and E. Witten. Vector bundles and F-theory. *Comm. Math. Phys.*, 187(3):679–743, 1997.

[FMW98] R. Friedman, J. Morgan, and E. Witten. Principal G-bundles over elliptic curves. *Math. Res. Lett.*, 5(1-2):97–118, 1998.

[FMW99] R. Friedman, J. Morgan, and E. Witten. Vector bundles over elliptic fibrations. *J. Algebraic Geom.*, 8(2):279–401, 1999.

[Gab81] O. Gabber. Some theorems on Azumaya algebras. In *The Brauer group (Sem., Les Plans-sur-Bex, 1980)*, pages 129–209. Springer, Berlin, 1981.

[Gir71] J. Giraud. *Cohomologie non abélienne*. Springer-Verlag, Berlin, 1971. Die Grundlehren der mathematischen Wissenschaften, Band 179.

[GMS00] O. Ganor, A. Mikhailov, and N. Saulina. Constructions of noncommutative instantons on T^4 and K_3. *Nuclear Phys. B*, 591(1-2):547–583, 2000.

[Gra91] A. Grassi. On minimal models of elliptic threefolds. *Math. Ann.*, 290(2):287–301, 1991.

[Gro68a] A. Grothendieck. Le groupe de Brauer. I,. Algèbres d'Azumaya et interprétations diverses. In *Dix Exposés sur la Cohomologie des Schémas*, pages 46–66. North-Holland, Amsterdam, 1968.

[Gro68b] A. Grothendieck. Le groupe de Brauer. II. Théorie cohomologique. In *Dix Exposés sur la Cohomologie des Schémas*, pages 67–87. North-Holland, Amsterdam, 1968.

[Gro68c] A. Grothendieck. Le groupe de Brauer. III. Exemples et compléments. In *Dix Exposés sur la Cohomologie des Schémas*, pages 88–188. North-Holland, Amsterdam, 1968.

[Gro98] M. Gross. Special Lagrangian fibrations. I. Topology. In *Integrable systems and algebraic geometry (Kobe/Kyoto, 1997)*, pages 156–193. World Sci. Publishing, River Edge, NJ, 1998.

[Gro99] M. Gross. Special Lagrangian fibrations. II. Geometry. A survey of techniques in the study of special Lagrangian fibrations. In *Surveys in differential geometry: differential geometry inspired by string theory*, volume 5 of *Surv. Differ. Geom.*, pages 341–403. Int. Press, Boston, MA, 1999.

[Har66] R. Hartshorne. Residues and duality. Lecture Notes in Mathematics, No. 20, Springer-Verlag, 1966.

[Hit01] N. Hitchin. Lectures on special Lagrangian submanifolds. In *Winter School on Mirror Symmetry, Vector Bundles and Lagrangian Submanifolds (Cambridge, MA, 1999)*, volume 23 of *AMS/IP Stud. Adv. Math.*, pages 151–182. Amer. Math. Soc., Providence, RI, 2001.

[Hoo82] R. Hoobler. When is $\mathrm{Br}(X) = \mathrm{Br}'(X)$? In *Brauer groups in ring theory and algebraic geometry (Wilrijk, 1981)*, pages 231–244. Springer, Berlin, 1982.

[HSc05] D. Huybrechts and S. Schroeer. The Brauer group of analytic $K3$ surfaces, Int. Math. Res. Not. 2003, no. 50, 2687–2698.

[HSt04a] D. Huybrechts and P. Stellari. Equivalences of twisted $K3$ surfaces, Math. Ann. 332 (2005), no. 4, 901–936.

[HSt04b] D. Huybrechts and P. Stellari. Proof of Căldăraru's conjecture. An appendix to a paper by Yoshioka, 2004, math.AG/0411541.

[IN99] Y. Ito and I. Nakamura. Hilbert schemes and simple singularities. In *New trends in algebraic geometry (Warwick, 1996)*, volume 264 of *London Math. Soc. Lecture Note Ser.*, pages 151–233. Cambridge Univ. Press, Cambridge, 1999.

[Kaw02] Yujiro Kawamata. D-equivalence and K-equivalence, J. Differential Geom. **61** (2002), no. 1, 147–171.

[KKO01] A. Kapustin, A. Kuznetsov, and D. Orlov. Noncommutative instantons and twistor transform. *Comm. Math. Phys.*, 221(2):385–432, 2001.

[Kod63] K. Kodaira. On compact analytic surfaces. II, III. *Ann. of Math. (2) 77 (1963), 563–626; ibid.*, 78:1–40, 1963.

[Kon01] M. Kontsevich. Deformation quantization of algebraic varieties. *Lett. Math. Phys.*, 56(3):271–294, 2001. EuroConférence Moshé Flato 2000, Part III (Dijon).

[LMB00] G. Laumon and L. Moret-Bailly. *Champs algébriques*. Springer-Verlag, Berlin, 2000.

[Mil80] J. Milne. *Etale cohomology*. Princeton University Press, Princeton, N.J., 1980.

[Mir83] R. Miranda. Smooth models for elliptic threefolds. In *The birational geometry of degenerations (Cambridge, Mass., 1981)*, pages 85–133. Birkhäuser Boston, Mass., 1983.

[Muk81] S. Mukai. Duality between $D(X)$ and $D(\hat{X})$ with its application to Picard sheaves. *Nagoya Math. J.*, 81:153–175, 1981.

[Nak02] N. Nakayama. Global structure of an elliptic fibration, Publ. Res. Inst. Math. Sci. 38 (2002), no. 3, 451–649.

[NS98] N. Nekrasov and A. Schwarz. Instantons on noncommutative $\mathbb{R}^4$, and $(2,0)$ superconformal six-dimensional theory. *Comm. Math. Phys.*, 198(3):689–703, 1998.

[Orl97] D. Orlov. Equivalences of derived categories and $K3$ surfaces. *J. Math. Sci. (New York)*, 84(5):1361–1381, 1997. Algebraic geometry, 7.

[Orl02] D. Orlov. Derived categories of coherent sheaves on abelian varieties and equivalences between them. *Izv. Ross. Akad. Nauk Ser. Mat.*, 66(3):131–158, 2002.

[Pol96] A. Polishchuk. Symplectic biextensions and a generalization of the Fourier-Mukai transform. *Math. Res. Lett.*, 3(6):813–828, 1996.

[Pol02] A. Polishchuk. Analogue of Weil representation for abelian schemes. *J. Reine Angew. Math.*, 543:1–37, 2002.

[Ray70] M. Raynaud. Spécialisation du foncteur de Picard. *Inst. Hautes Études Sci. Publ. Math.*, 38:27–76, 1970.

[Sch01] S. Schröer. There are enough Azumaya algebras on surfaces. *Math. Ann.*, 321(2):439–454, 2001.

[SGA7-I] Groupes de monodromie en géométrie algébrique. I. Lecture Notes in Mathematics, Vol. 288, Springer, 1972. Séminaire de Géométrie Algébrique du Bois-Marie 1967–1969 (SGA 7 I), Dirigé par A. Grothendieck. Avec la collaboration de M. Raynaud et D. S. Rim.

[ST01] P. Seidel and R. Thomas. Braid group actions on derived categories of coherent sheaves. *Duke Math. J.*, 108(1):37–108, 2001.

[SYZ96] A. Strominger, S.-T. Yau, and E. Zaslow. Mirror symmetry is T-duality. *Nuclear Phys. B*, 479(1-2):243–259, 1996.

Editorial Information

To be published in the *Memoirs*, a paper must be correct, new, nontrivial, and significant. Further, it must be well written and of interest to a substantial number of mathematicians. Piecemeal results, such as an inconclusive step toward an unproved major theorem or a minor variation on a known result, are in general not acceptable for publication.

Papers appearing in *Memoirs* are generally at least 80 and not more than 200 published pages in length. Papers less than 80 or more than 200 published pages require the approval of the Managing Editor of the Transactions/Memoirs Editorial Board.

As of January 31, 2008, the backlog for this journal was approximately 17 volumes. This estimate is the result of dividing the number of manuscripts for this journal in the Providence office that have not yet gone to the printer on the above date by the average number of monographs per volume over the previous twelve months, reduced by the number of volumes published in four months (the time necessary for preparing a volume for the printer). (There are 6 volumes per year, each usually containing at least 4 numbers.)

A Consent to Publish and Copyright Agreement is required before a paper will be published in the *Memoirs*. After a paper is accepted for publication, the Providence office will send a Consent to Publish and Copyright Agreement to all authors of the paper. By submitting a paper to the *Memoirs*, authors certify that the results have not been submitted to nor are they under consideration for publication by another journal, conference proceedings, or similar publication.

Information for Authors

Memoirs are printed from camera copy fully prepared by the author. This means that the finished book will look exactly like the copy submitted.

Initial submission. The AMS uses Centralized Manuscript Processing for initial submissions. Authors should submit a PDF file using the Initial Manuscript Submission form found at `www.ams.org/cgi-bin/peertrack/submission.pl`, or send one copy of the manuscript to the following address: Centralized Manuscript Processing, MEMOIRS OF THE AMS, 201 Charles Street, Providence, RI 02904-2294 USA. If a paper copy is being forwarded to the AMS, indicate that it is for it Memoirs and include the name of the corresponding author, contact information such as email address or mailing address, and the name of an appropriate Editor to review the paper (see the list of Editors below).

The paper must contain a *descriptive title* and an *abstract* that summarizes the article in language suitable for workers in the general field (algebra, analysis, etc.). The *descriptive title* should be short, but informative; useless or vague phrases such as "some remarks about" or "concerning" should be avoided. The *abstract* should be at least one complete sentence, and at most 300 words. Included with the footnotes to the paper should be the 2000 *Mathematics Subject Classification* representing the primary and secondary subjects of the article. The classifications are accessible from `www.ams.org/msc/`. The list of classifications is also available in print starting with the 1999 annual index of *Mathematical Reviews*. The Mathematics Subject Classification footnote may be followed by a list of *key words and phrases* describing the subject matter of the article and taken from it. Journal abbreviations used in bibliographies are listed in the latest *Mathematical Reviews* annual index. The series abbreviations are also accessible from `www.ams.org/publications/`. To help in preparing and verifying references, the AMS offers MR Lookup, a Reference Tool for Linking, at `www.ams.org/mrlookup/`.

Electronically prepared manuscripts. The AMS encourages electronically prepared manuscripts, with a strong preference for AMS-LaTeX. To this end, the Society has prepared AMS-LaTeX author packages for each AMS publication. Author packages include instructions for preparing electronic manuscripts, samples, and a style file that generates the particular design specifications of that publication series. Though $\mathcal{AMS}$-LaTeX is the highly preferred format of TeX, author packages are also available in $\mathcal{AMS}$-TeX.

Authors may retrieve an author package from the AMS website starting from `www.ams.org/tex/` or via FTP to `ftp.ams.org` (login as `anonymous`, enter username as password, and type `cd pub/author-info`). The *AMS Author Handbook* and the *Instruction Manual* are available in PDF format following the author packages link from `www.ams.org/tex/`. The author package can also be obtained free of charge by sending email to `tech-support@ams.org` (Internet) or from the Publication Division, American Mathematical Society, 201 Charles St., Providence, RI 02904-2294, USA. When requesting an author package, please specify $\mathcal{AMS}$-LaTeX or $\mathcal{AMS}$-TeX and the publication in which your paper will appear. Please be sure to include your complete mailing address.

After acceptance. The final version of the electronic file should be sent to the Providence office (this includes any TeX source file, any graphics files, and the DVI or PostScript file) immediately after the paper has been accepted for publication.

Before sending the source file, be sure you have proofread your paper carefully. The files you send must be the EXACT files used to generate the proof copy that was accepted for publication. For all publications, authors are required to send a printed copy of their paper, which exactly matches the copy approved for publication, along with any graphics that will appear in the paper.

Accepted electronically prepared files can be submitted via the web at `www.ams.org/submit-book-journal/`, sent via FTP, or sent on CD-Rom or diskette to the Electronic Prepress Department, American Mathematical Society, 201 Charles Street, Providence, RI 02904-2294 USA. TeX source files, DVI files, and PostScript files can be transferred over the Internet by FTP to the Internet node `ftp.ams.org` (130.44.1.100). When sending a manuscript electronically via CD-Rom or diskette, please be sure to include a message identifying the paper as a Memoir.

Electronically prepared manuscripts can also be sent via email to `pub-submit@ams.org` (Internet). In order to send files via email, they must be encoded properly. (DVI files are binary and PostScript files tend to be very large.)

Electronic graphics. Comprehensive instructions on preparing graphics are available at `www.ams.org/jourhtml/`. A few of the major requirements are given here.

Submit files for graphics as EPS (Encapsulated PostScript) files. This includes graphics originated via a graphics application as well as scanned photographs or other computer-generated images. If this is not possible, TIFF files are acceptable as long as they can be opened in Adobe Photoshop or Illustrator. No matter what method was used to produce the graphic, it is necessary to provide a paper copy to the AMS.

Authors using graphics packages for the creation of electronic art should also avoid the use of any lines thinner than 0.5 points in width. Many graphics packages allow the user to specify a "hairline" for a very thin line. Hairlines often look acceptable when proofed on a typical laser printer. However, when produced on a high-resolution laser imagesetter, hairlines become nearly invisible and will be lost entirely in the final printing process.

Screens should be set to values between 15% and 85%. Screens which fall outside of this range are too light or too dark to print correctly. Variations of screens within a graphic should be no less than 10%.

Inquiries. Any inquiries concerning a paper that has been accepted for publication should be sent to `memo-query@ams.org` or directly to the Electronic Prepress Department, American Mathematical Society, 201 Charles St., Providence, RI 02904-2294 USA.

Editors

This journal is designed particularly for long research papers, normally at least 80 pages in length, and groups of cognate papers in pure and applied mathematics. Papers intended for publication in the *Memoirs* should be addressed to one of the following editors. The AMS uses Centralized Manuscript Processing for initial submissions to AMS journals. Authors should follow instructions listed on the Initial Submission page found at www.ams.org/memo/memosubmit.html.

Algebra to ALEXANDER KLESHCHEV, Department of Mathematics, University of Oregon, Eugene, OR 97403-1222; email: ams@noether.uoregon.edu

Algebraic geometry and its application to MINA TEICHER, Emmy Noether Research Institute for Mathematics, Bar-Ilan University, Ramat-Gan 52900, Israel; email: teicher@macs.biu.ac.il

Algebraic geometry to DAN ABRAMOVICH, Department of Mathematics, Brown University, Box 1917, Providence, RI 02912; email: amsedit@math.brown.edu

Algebraic number theory to V. KUMAR MURTY, Department of Mathematics, University of Toronto, 100 St. George Street, Toronto, ON M5S 1A1, Canada; email: murty@math.toronto.edu

Algebraic topology to ALEJANDRO ADEM, Department of Mathematics, University of British Columbia, Room 121, 1984 Mathematics Road, Vancouver, British Columbia, Canada V6T 1Z2; email: adem@math.ubc.ca

Combinatorics to JOHN R. STEMBRIDGE, Department of Mathematics, University of Michigan, Ann Arbor, Michigan 48109-1109; email: FRS@umich.edu

Complex analysis and harmonic analysis to ALEXANDER NAGEL, Department of Mathematics, University of Wisconsin, 480 Lincoln Drive, Madison, WI 53706-1313; email: nagel@math.wisc.edu

Differential geometry and global analysis to LISA C. JEFFREY, Department of Mathematics, University of Toronto, 100 St. George St., Toronto, ON Canada M5S 3G3; email: jeffrey@math.toronto.edu

Functional analysis and operator algebras to DIMITRI SHLYAKHTENKO, Department of Mathematics, University of California, Los Angeles, CA 90095; email: shlyakht@math.ucla.edu

Geometric analysis to WILLIAM P. MINICOZZI II, Department of Mathematics, Johns Hopkins University, 3400 N. Charles St., Baltimore, MD 21218; email: trans@math.jhu.edu

Geometric analysis to MARK FEIGHN, Math Department, Rutgers University, Newark, NJ 07102; email: feighn@andromeda.rutgers.edu

Harmonic analysis, representation theory, and Lie theory to ROBERT J. STANTON, Department of Mathematics, The Ohio State University, 231 West 18th Avenue, Columbus, OH 43210-1174; email: stanton@math.ohio-state.edu

Logic to STEFFEN LEMPP, Department of Mathematics, University of Wisconsin, 480 Lincoln Drive, Madison, Wisconsin 53706-1388; email: lempp@math.wisc.edu

Number theory to JONATHAN ROGAWSKI, Department of Mathematics, University of California, Los Angeles, CA 90095; email: jonr@math.ucla.edu

Partial differential equations to GUSTAVO PONCE, Department of Mathematics, South Hall, Room 6607, University of California, Santa Barbara, CA 93106; email: ponce@math.ucsb.edu

Partial differential equations and dynamical systems to PETER POLACIK, School of Mathematics, University of Minnesota, Minneapolis, MN 55455; email: polacik@math.umn.edu

Probability and statistics to RICHARD BASS, Department of Mathematics, University of Connecticut, Storrs, CT 06269-3009; email: bass@math.uconn.edu

Real analysis and partial differential equations to DANIEL TATARU, Department of Mathematics, University of California, Berkeley, Berkeley, CA 94720; email: tataru@math.berkeley.edu

All other communications to the editors should be addressed to the Managing Editor, ROBERT GURALNICK, Department of Mathematics, University of Southern California, Los Angeles, CA 90089-1113; email: guralnic@math.usc.edu.

Titles in This Series

905 **Dominic Verity,** Complicial sets characterising the simplicial nerves of strict ω-categories, 2008

904 **William M. Goldman and Eugene Z. Xia,** Rank one Higgs bundles and representations of fundamental groups of Riemann surfaces, 2008

903 **Gail Letzter,** Invariant differential operators for quantum symmetric spaces, 2008

902 **Bertrand Toën and Gabriele Vezzosi,** Homotopical algebraic geometry II: Geometric stacks and applications, 2008

901 **Ron Donagi and Tony Pantev (with an appendix by Dmitry Arinkin),** Torus fibrations, gerbes, and duality, 2008

900 **Wolfgang Bertram,** Differential geometry, Lie groups and symmetric spaces over general base fields and rings, 2008

899 **Piotr Hajłasz, Tadeusz Iwaniec, Jan Malý, and Jani Onninen,** Weakly differentiable mappings between manifolds, 2008

898 **John Rognes,** Galois extensions of structured ring spectra/Stably dualizable groups, 2008

897 **Michael I. Ganzburg,** Limit theorems of polynomial approximation with exponential weights, 2008

896 **Michael Kapovich, Bernhard Leeb, and John J. Millson,** The generalized triangle inequalities in symmetric spaces and buildings with applications to algebra, 2008

895 **Steffen Roch,** Finite sections of band-dominated operators, 2008

894 **Martin Dindoš,** Hardy spaces and potential theory on C^1 domains in Riemannian manifolds, 2008

893 **Tadeusz Iwaniec and Gaven Martin,** The Beltrami Equation, 2008

892 **Jim Agler, John Harland, and Benjamin J. Raphael,** Classical function theory, operator dilation theory, and machine computation on multiply-connected domains, 2008

891 **John H. Hubbard and Peter Papadopol,** Newton's method applied to two quadratic equations in $\mathbb{C}^2$ viewed as a global dynamical system, 2008

890 **Steven Dale Cutkosky,** Toroidalization of dominant morphisms of 3-folds, 2007

889 **Michael Sever,** Distribution solutions of nonlinear systems of conservation laws, 2007

888 **Roger Chalkley,** Basic global relative invariants for nonlinear differential equations, 2007

887 **Charlotte Wahl,** Noncommutative Maslov index and eta-forms, 2007

886 **Robert M. Guralnick and John Shareshian,** Symmetric and alternating groups as monodromy groups of Riemann surfaces I: Generic covers and covers with many branch points, 2007

885 **Jae Choon Cha,** The structure of the rational concordance group of knots, 2007

884 **Dan Haran, Moshe Jarden, and Florian Pop,** Projective group structures as absolute Galois structures with block approximation, 2007

883 **Apostolos Beligiannis and Idun Reiten,** Homological and homotopical aspects of torsion theories, 2007

882 **Lars Inge Hedberg and Yuri Netrusov,** An axiomatic approach to function spaces, spec tral synthesis and Luzin approximation, 2007

881 **Tao Mei,** Operator valued Hardy spaces, 2007

880 **Bruce C. Berndt, Geumlan Choi, Youn-Seo Choi, Heekyoung Hahn, Boon Pin Yeap, Ae Ja Yee, Hamza Yesilyurt, and Jinhee Yi,** Ramanujan's forty identities for Rogers-Ramanujan functions, 2007

879 **O. García-Prada, P. B. Gothen, and V. Muñoz,** Betti numbers of the moduli space of rank 3 parabolic Higgs bundles, 2007

878 **Alessandra Celletti and Luigi Chierchia,** KAM stability and celestial mechanics, 2007

877 **María J. Carro, José A. Raposo, and Javier Soria,** Recent developments in the theory of Lorentz spaces and weighted inequalities, 2007

For a complete list of titles in this series, visit the AMS Bookstore at **www.ams.org/bookstore/**.